NATURE CRIME

NATURE CRIME

HOW WE'RE GETTING CONSERVATION WRONG

ROSALEEN DUFFY

YALE UNIVERSITY PRESS
NEW HAVEN AND LONDON

For information about this and other Yale University Press publications, please contact:

U.S. Office: sales.press@yale.edu   yalepress.yale.edu
Europe Office: sales@yaleup.co.uk   www.yalebooks.co.uk

Set in Janson Text by IDSUK (DataConnection) Ltd
Printed in Great Britain by TJ International Ltd, Padstow, Cornwall

Library of Congress Cataloging-in-Publication Data

Duffy, Rosaleen.
    Nature crime: how we're getting conservation wrong / Rosaleen Duffy.
        p. cm.
    ISBN 978-0-300-15434-4 (hardback: alk. paper) 1. Nature conservation.
2. Environmental responsibility. 3. Nature—Effect of human beings on.
4. Endangered species. I. Title.
    QH75.D838 2010
    333.95'416—dc22

                                                                2010008937

A catalogue record for this book is available from the British Library.

10 9 8 7 6 5 4 3 2 1
2014 2013 2012 2011 2010

CONTENTS

Acknowledgements   vi

List of Illustrations   ix

List of Abbreviations   xi

Introduction   1

Chapter One: The International Wildlife Trade   15

Chapter Two: Global Action, Local Costs   45

Chapter Three: Wildlife Wars: Poaching and   79
                  Anti-Poaching

Chapter Four: Rhino Horn, Ivory and the Trade Ban   113
                  Controversy

Chapter Five: Guerrillas to Gorillas: Blood Diamonds   155
                  and Coltan

Chapter Six: Tourist Saviours   187

Conclusion   217

Notes   225

Bibliography   243

Index   253

ACKNOWLEDGEMENTS

This book draws on over fifteen years of research, during which time I have had the privilege to meet and interact with conservation professionals, representatives of local communities, tour operators, journalists and government officials. Each has influenced this book, because their thoughts and arguments have allowed me to develop the argument and case material it contains; I could not have written this book without them. That said, any errors are my responsibility.

Unfortunately I cannot mention all the people I have interviewed over the years, but there are a few people who shaped my thinking at critical junctures: Professor Marshall Murphree and Dr Vupenyu Dzingirai at the University of Zimbabwe; Dr Rowan Martin and Dr Brian Child (formerly of the Zimbabwe Parks Department, now at the University of Florida); Richard Thomas at TRAFFIC-International; Inspector Brian Stuart at the National Wildlife Crime Unit; Tom Fazakerly and Bob Hodgson in Lancaster; Nivo Ravelojaona of Za Tour Madagascar, Norosoa

Acknowledgements

Ranivomboahangy at the Christina Dodwell Trust/Mitondrasoa in Madagascar; Steve Goodman, WWF-Madagascar; Dr Joseph Mbaiwa in the Okavango Research Centre, University of Botswana; Brian Gonzales at Traffic-Asia; Pio Coc of the Toledo Maya Cultural Council; Mary Vasquez in Belize; and Professor Christopher Clapham at Cambridge University. There are many other people who have provided me with critically important information but who prefer to remain anonymous; I still thank them for their courage in speaking out.

The research was also made possible through generous financial support from the UK Economic and Social Research Council (ESRC), consisting of grants for three separate projects: 'Neoliberalising Nature? A Comparative Analysis of Asian and African Elephant Based Ecotourism' (2007–8), with Lorraine Moore; 'Global Environmental Governance and Local Resistances' (2003–4); and 'The Geopolitics of Bioregions: Conservation and Erosion of National Boundaries' (1999–2000).

During the process of writing the book I have benefited from interacting with colleagues at Manchester University, within both the Politics department and the interdisciplinary Society and Environment Research Group (SERG). I would like to thank the conservation reading group for providing a stimulating space for debate, and for their comments on early drafts of parts of this book. I particularly thank Dr Daniel Brockington for his support – which was both intellectual and practical, especially during the final stages of writing. I also thank the Cambridge Geography reading group who read and discussed an early draft chapter with me. I am grateful to Desna MacKenzie, Mary Duffy, Professor Noel Castree, Dr Tim Forsyth, Dr Lorraine Moore, Dr Danielle Beswick, Professor Bill Adams and Dr Chris Sandbrook for their insightful comments on draft chapters. Equally, I thank the anonymous referees on this manuscript for their helpful comments, which challenged me to think carefully about the arguments it contained. I cannot thank my editor, Phoebe Clapham, enough; she was extremely patient while I wrote this book and gave me excellent feedback on the chapters as I wrote them.

Finally, I would like to thank the friends who have listened to me talk incessantly about wildlife while writing this book: David and Barbara Denver, Patrick Bishop, Graham, Sarah and Henry Smith, Jean and Olly Brown, Tracy, John and Corey Sartin, Vicky Mason and Kynan Gentry. Thanks also go to my family, especially Mum and Dad, who helped in so many different ways, as only parents can; Carmel, Simon, Peter, Joanna, Thomas and Ruby have all provided fun and walks – Morecambe Bay is a favourite, and some of the photos in this book were taken there. My brothers and sisters, nieces and nephews have supported me in lots of different ways – I hope my obsessions haven't put them off wildlife forever. I also want to thank Roisin and Gerry Cochrane, who have always been supportive of my diverse endeavours. Finally, this book is dedicated to my husband, Feargal Cochrane. As a fellow academic he understood the challenges of writing a book, and was a willing intellectual sounding board. More importantly, Feargal has been a patient and unwavering source of support, and has even put up with visiting me on fieldwork.

LIST OF ILLUSTRATIONS

All photographs taken by author

p. 16   1.1   Collecting birds' nests, Krabi, Thailand (2008)

p. 20   1.2   Collecting cicadas for food, Maerim Valley, Northern Thailand (2008)

p. 32   1.3   Pufferfish for sale to tourists, Phuket (2008)

p. 38   1.4   Morecambe Bay, UK (2010)

p. 40   1.5   Abandoned vehicle used to transport cockles to the shoreline (2010)

p. 42   1.6   Notice that the cockling beds are closed (2009)

p. 54   2.1   Clear demarcation of park boundary and agricultural land, Bwindi Impenetrable National Park, Uganda (2007)

p. 55   2.2   Elephant watching in the Kruger National Park 'wilderness' (2006)

p. 98   3.1   Community-based tourism scheme: polers and canoes in the Okavango Delta, Botswana (2008)

p. 108 3.2 Beach fenced off for tourism, Lake Victoria, Uganda (2007)

p. 124 4.1 Ivory ornaments at a Buddhist temple in Sri Lanka (2003)

p. 125 4.2 Ivory carving at the UN headquarters in New York, a gift from the People's Republic of China (2006)

p. 132 4.3 Flags to scare elephants, Okavango Delta, Botswana (2008)

p. 140 4.4 Ivory on sale illegally in New York (2008)

p. 151 4.5 Ivory on sale illegally in Ethiopia (2002)

p. 157 5.1 Welcoming tourists to the Parc National des Volcans, Rwanda (2007)

p. 159 5.2 Mountain gorilla, Parc National des Volcans in Rwanda (2007)

p. 170 5.3 UNHCR Nyakabanda camp for refugees from DRC (2007)

p. 177 5.4 Verreaux sifaka, Isalo National Park, Madagascar (2004)

p. 180 5.5 International sapphire dealers, Ilakaka, Madagascar (2004)

p. 191 6.1 Poster highlighting the plight of homeless hermit crabs (2007)

p. 193 6.2 Shell souvenirs for tourists, Thailand (2008)

p. 198 6.3 Wooden carvings for tourists, Cape Verde (2010)

p. 199 6.4 Seafood restaurant, Phuket, Thailand (2008)

p. 203 6.5 Marmoset offered for tourists to photograph, Playa del Carmen, Mexico (2009)

p. 208 6.6 Beach building, Playa del Carmen, Mexico (2009)

p. 209 6.7 Turtle nest marker, Akumal beach, Mexico (2009)

LIST OF ABBREVIATIONS

ACAP	Active Conservation Awareness Programme
ADFL	Alliance of Democratic Forces for the Liberation of Congo
ANGAP	Association Nationale pour la Gestion des Aires Protégées
ARRC	African Rainforest and River Conservation
AWF	African Wildlife Foundation
CAMPFIRE	Communal Areas Management Programme for Indigenous Resources
CBD	Convention on Biological Diversity
CBNRM	Community-Based Natural Resource Management
CI	Conservation International
CITES	Convention on the International Trade in Endangered Species
DEFRA	Department for Environment, Food and Rural Affairs
DRC	Democratic Republic of Congo
DSWF	David Shepherd Wildlife Foundation
EIA	Environmental Investigation Agency
GLTP	Great Limpopo Transfrontier Park
HSUS	Humane Society of the US
ICCN	Institut Congolais pour la Conservation de la Nature/Congolese Institute for Nature Conservation
ICDP	Integrated Conservation and Development Project
IDPs	Internally Displaced People
IFAW	International Fund for Animal Welfare
ITRG	Ivory Trade Review Group

IUCN	International Union for the Conservation of Nature
IUCN-SSC	IUCN Species Survival Commission
KWS	Kenya Wildlife Services
MRAG	Marine Resources Assessment Group
NGO	non-governmental organization
NTCA	National Tiger Conservation Authority
NWCU	National Wildlife Crime Unit
PAC	Problem Animal Control
PAW	Partnership for Action Against Wildlife Crime
PHASA	Professional Hunters Association of South Africa
QMM	Qit-Fer Minerals Madagascar
RPF	Rwandan Patriotic Front
RSPB	Royal Society for the Protection of Birds
RSPCA	Royal Society for the Prevention of Cruelty to Animals
SADF	South African Defence Force
SIDA	Swedish International Development Cooperation Agency
TCM	Traditional Chinese Medicine
TFCAs	Transfrontier Conservation Areas
TNC	The Nature Conservancy
TRAFFIC	Trade Records and Analysis of Flora and Fauna in Commerce
UNITA	União Nacional para a Independência Total de Angola/ The National Union for the Total Independence of Angola
UNODC	United Nations Office on Drugs and Crime
USAID	United States Agency for International Development
WCS	Wildlife Conservation Society
WWF	World Wide Fund for Nature

For Feargal

INTRODUCTION

My home is vanishing. The forests I once ruled are being cut down, and my hunting grounds destroyed. Will you remember me? Will you show your grandchildren pictures of the beautiful cats that used to share your planet? Will they believe that such magnificent creatures were allowed to die out? All of us, finally and forever. It's not too late. But soon it will be. *Don't let me go.*

THIS WORLD WIDE FUND for Nature fundraising leaflet for tiger conservation was circulated in November 2009. It neatly communicates the problem: wildlife is under threat and we need to act urgently. Conservation non-governmental organizations (NGOs) like World Wide Fund for Nature (WWF) and Conservation International tell us that the loss of wildlife is one of the most important challenges facing our planet, that we are facing an extinction crisis to rival the end of the dinosaurs. We fear that some of the world's most iconic species will disappear from the earth forever, that future generations will not be able to see mountain gorillas, pandas, tigers or elephants in the wild, if at all.

The International Union for the Conservation of Nature (IUCN)'s *Red List* tells us that 360 vertebrates, 373 invertebrates and 110 plants have become extinct, or extinct in the wild, since AD 1500.[1] Fears about species loss have been stoked by high-profile public figures, such as Paul Ehrlich and Edward O. Wilson, who argue that the loss of tropical forests alone is eliminating anything from 4,000 to 40,000 species per year.[2] Some scientists have even argued that we are living through another great extinction crisis – the end of the dinosaurs was the fifth extinction, but this sixth extinction is the result of human interference in the natural world.[3] This raises the question of why extinctions matter. In many writings on conservation it is simply assumed that extinction is a 'bad thing' that needs to be tackled urgently. Writers such as Norman Myers and Edward O. Wilson point to the potentially harmful impacts on ecosystems and on human life, arguing that extinctions matter because they are indicators of the wider health of the planet, which sustains human life. If too many species go extinct them we risk ecosystem collapses. Norman Myers puts it bluntly: 'we are dealing with a matter of exceptional importance and unique urgency . . . There is a need to address the extinction of populations, which are the main providers of ecosystem functions and services.'[4]

Concerns about species and habitat loss have motivated governments, individuals and private companies to support wildlife conservation around the world, and, judging by the amount of money raised and effort expended, conservation is big business. Many of us are loyal members of conservation organizations: we donate, adopt or sponsor animals, and support their campaigns. For example, WWF-UK has 168,417 members, and 250 major donors.[5] WWF-US has 1.2 million members in the United States and nearly five million supporters worldwide.[6] It is difficult to establish precisely how much money is spent each year on conservation, but the four biggest NGOs – Conservation International, the Wildlife Conservation Society, WWF and the Nature Conservancy (TNC) – invested a total of US$490 million in conservation in the developing world in 2002, with a similar amount being spent by other, smaller organizations.[7] TNC claims to be the world's largest conservation organization with net assets worth US$4.3 billion, and, like other big conservation NGOs, it has links with a growing list of corporate partners, including AT&T, Hewlett Packard, Avon, Oracle, Cisco, BP and General Motors.[8] Save the Tiger Fund has benefited from one of the largest corporate donations in history: ExxonMobil provides US$1 million annually, and has donated more than US$13 million in total for tiger-range countries since 1992.[9] The role of NGOs will be discussed in more detail in Chapter 2, but these examples give some indication of their growing power: through funding for specific forms of conservation they have the ability to project their vision across the world.

Considerable effort, funding and attention has therefore been lavished on some species and the areas they live in. Tigers, gorillas, elephants, pandas and rhinos spring to mind, although there are many more species that cause concern to conservationists around the world. We are accustomed to bad news stories in conservation, but there have also been some important successes. Some species have been brought back from the brink of extinction through careful management and investment. For example, the number of white rhinos has increased from only 50–200 at the start of the twentieth century to about 17,500 today.[10] In the UK the Royal Society for the Protection of Birds (RSPB) has successfully reintroduced sea eagles to Scotland, where they became extinct in 1916; a viable population has now been established, and in 2009 there were forty-three breeding pairs that successfully reared thirty-six chicks.[11] The Durrell Wildlife Conservation Trust has also succeeded in saving the Mauritius kestrel from extinction, and its captive breeding programme for the Mauritius pink pigeon has enabled twenty pigeons to be returned to Mauritius, as a result of which the species is now listed as 'endangered' rather than 'critically endangered' by the IUCN.[12] These are important achievements in conservation, and they should not be overlooked as we rush to declare an extinction crisis. These successes indicate that it is possible for us to save individual species, even when the outlook seems bleak.

However, it is foolish to assume that these success stories mean that conservationists do not have major challenges on their hands. In 1970 there were approximately 65,000 black rhinos in the wild in Africa, but by 1992 there were just 2,300, a staggering 96 per cent decrease.[13] In the late 1980s elephant populations crashed from 1.3 million to just 600,000 due to poaching for ivory. WWF estimates that elephant range declined from 7.3 million square kilometres in 1979 to 5.9 million in 1987 and to 5.7 million by 1998; of the remaining range area available for wild elephants today, almost 80 per cent falls outside protected areas.[14] And IUCN figures show that tigers have lost a massive 93 per cent of their historical range, and that there are just 5,000 left in the wild.[15] Clearly some species are facing serious threats to their long-term survival.

Since habitat is crucial in sustaining animal populations, conservation efforts are generally focused around the development and management of protected areas – that is, national parks and reserves, where wildlife can thrive, undisturbed by human activity. This is not a new idea; some of the world's most famous parks, such as Yellowstone and Yosemite in the United States of

America, were established in the late nineteenth century. This model of the people-free wilderness was exported by European colonial administrations, which set up new networks of protected areas in Africa and Asia in the nineteenth and twentieth centuries. However, the process of creating new protected areas did not end with the dying days of colonialism. The number of protected areas around the world increased substantially between 1985 and 1995, and, while this spectacular growth has waned in the last few years, new protected areas are still springing up all over the world.[16] For the conservation movement this is a cause for celebration, because it means that more and more areas and species are sheltered from the harmful effects of human development.

However, as I shall discuss in this book, these 'successes' increasingly come at a cost. The development of two new protected areas, in the Chagos Archipelago and Southern Madagascar, illustrate some of the problems involved.

In 2010 the UK Foreign and Commonwealth Office began a consultation process for developing the world's largest marine protected area in Chagos Archipelago, in the Indian Ocean. The area is known for its pristine coral reefs, and the scheme is supported by a powerful network of conservation organizations including Kew Gardens, London Zoo, the RSPB, the Royal Society and the Marine Conservation Society. However, the islands' picture-perfect image hides a darker history. Their residents were forcibly evicted in the late 1960s by the British government, to make way for the development of a US military base on Diego Garcia, the largest of the islands. Since then, more than 4,000 evictees have fought a legal battle to return to their homeland, taking their case through the British courts and finally to the European Court of Human Rights. They are concerned that the new marine conservation area is designed to 'greenwash' the existing military base and cement the islanders' exile; even if they were allowed to return home, the proposed ban on fishing would mean that they could not use local resources for subsistence or income generation.[17] So the Chagossians would continue to be excluded by the creation of the protected area, while the US military would not.[18] IUCN has endorsed the plan, despite criticisms from its own legal advisors that it breaches the rights of evicted communities. The Chagos case draws our attention to the darker side of conservation: in particular, the power of international NGOs and the human costs of their 'good intentions' for wildlife.

Chagos is not an isolated problem. Conservation International, WWF and Wildlife Conservation Society have all been active in pushing forward the development of new protected areas in Madagascar since 2004 (discussed in chapter two). But this process has also been fraught with concerns about the displacement and marginalization of local communities. The case of a new ilmenite (titanium dioxide) mine by Qit-Fer Minerals Madagascar (QMM) in Fort Dauphin in southern Madagascar illustrates the major themes of this book: the importance of international conservation NGOs, the development of powerful new alliances with the private sector, the ways local communities are alienated by conservation schemes, and how conservation rules defending wildlife brand some people as criminals.

QMM is 80 per cent owned by the international mining corporation Rio Tinto and 20 per cent owned by the government of Madagascar. The Malagasy government is funding the construction of a port for the mine through its Integrated Growth Poles Project; ultimately funded by the World Bank, the project is intended to help provide the business environment to stimulate and lead economic growth in three selected regions of the country.[19] Rio Tinto claims that the mine project will deliver 'broader economic development in the country while providing conservation opportunities'.[20] The company has zoned areas around the mine which will become part of the national network of protected areas in Madagascar. It also set up an ecotourism project which has been running since 2000 to allow local communities to benefit from the conservation initiatives established by Rio Tinto.[21] QMM has worked closely with the Royal Botanical Gardens at Kew, Conservation International, Missouri Botanical Gardens, Earthwatch and the Smithsonian Institute to mitigate the social and environmental impacts of such a massive mining operation.[22] On the surface this looks like a good way to 'offset' the environmental damage caused by mining, and appears to provide genuine social and economic benefits to local communities.

But the mine has been heavily criticized by Friends of the Earth, which stated that the project threatens unique forest resources and would leave local people struggling to survive, despite promises of jobs and compensation. Local communities have complained that the compensation payments are not sufficient, since land prices have risen in that area, and that promises of employment have not materialized.[23] This reveals how well-intentioned conservation schemes can end up alienating local communities who have to live with the everyday costs of the new initiatives. Conservation does not constitute a neat win–win scenario, and often this is the result of the failure to recognize the complex politics involved in developing new initiatives.

This book draws on over fifteen years of research on conservation, during which time I have carried out numerous interviews with conservation professionals all over the world. I became accustomed to the puzzled look when they realised that my academic background was politics and international relations, not biology or environmental management. When I have ventured questions about the conflicts, arguments, divisions and wrangles over how to conserve a specific species, I have often been met with a dismissive comment: that it is all politics, it gets in the way of the science, of the real practice. Conservation professionals seem to find the 'politics' of it all rather unimportant and exhausting. My rejoinder is that politics is what *matters*: the inability to negotiate these conflicts and work with people on the ground is where conservation often sows the seeds of its own doom. The failure to recognize that their apparently neutral and science-based vision of conservation is a culturally embedded one, and that they are 'power actors' in the international system, lies at the heart of the problem.

While I have had a lifelong interest in wildlife and conservation, my formal research career began in Zimbabwe in 1994 where I completed fieldwork for my PhD. Since then I have been fortunate to research and experience conservation in action in Mexico, Belize, Madagascar, South Africa, Ethiopia, Thailand and Botswana. This book utilizes that ongoing fieldwork, which includes approximately 300 interviews with conservation policy makers, NGO representatives, scientists, journalists and government officials, as well as discussions with local communities and my own personal observations and experiences. Fieldwork material was supplemented with analysis of documents produced by NGOs, the World Bank, national governments and the wider writings of fellow academics committed to understanding the harmful effects of good intentions. The photographs in this book were all taken by me; they serve as illustrations and an attempt to place conservation in its social context. This presented a real ethical dilemma; I have tried to ensure that people are not made 'invisible' in the photographs, but equally have had to make sure that individuals, businesses and organizations were not identifiable, unless the photograph was taken as part of a public event.

This book examines the headline-grabbing stories of conservation, like ivory poaching, the gorilla murders of 2007 and the tiger bone trade, to work out how conservation can be made more effective. My intention is to improve our understanding of conservation by delving behind the assumptions about issues like poaching and the trade in wildlife. We often assume that 'people' are the enemies of wildlife conservation: they are the illegal traders, the poachers, the hunters and the habitat destroyers. Similarly we believe that individuals in conservation organizations and rangers in national parks are wildlife saviours who protect the planet's precious animals and environments from their enemies. However, these images are oversimplifications. This book will question the idea that there are wildlife enemies and wildlife saviours and explore how the good intentions of conservation end up alienating local communities. This is vital because failing to tackle such injustices damages wildlife conservation in the long run. This means we have to confront a series of challenging questions: why do global conservation policies not always achieve the desired effect? What are the impacts of protected areas on local communities? How important are international conservation NGOs? Why are there sharp divisions over the best ways to achieve conservation? Why is it that tourism has failed to be the magic bullet for wildlife conservation? Answering these questions requires us to see conservation in its global context: the bigger picture shapes the successes and failures of conservation on the ground.

One of the first issues tackled by this book is what is really driving the wildlife trade, and why it is so hard to control. The wildlife trade is one of the most valuable trades in the world. TRAFFIC-International estimates that in the early 1990s the value of legally traded wildlife products was US$160 billion per annum, legal wood exports were worth US$132 billion, and legal seafood exports were valued at US$50 billion. The value of the illegal trade in wildlife products is harder to estimate, but is likely to be worth between US$10 and US$20 billion per year, second only to the international drugs trade in the ranks of illegal exchange.[24] This trade has a massive impact on wildlife all around the world.

All sorts of different species are traded, not just ivory, rhino horn and tiger parts. All of us are consumers of wildlife in one way or another. We eat wildlife, we wear it as clothing and accessories, we consume it as medicine and we buy ornaments and souvenirs made from it. We might eat shellfish and caviar, wear shoes made from crocodile and snakeskin or buy ornaments made from tropical hardwoods. Our consumption habits also impact on wildlife because the goods we demand come from wildlife-rich areas. We wear diamond and sapphire jewellery made from stones mined in areas where gorillas or lemurs live; and we carry around mobile phones which use coltan, 80 per cent of which comes from a single national park in Democratic Republic of Congo. Through global consumer culture, our everyday lives are bound up with the fate of wildlife populations that are far distant from us.

In this book I will explore the patterns in the wildlife trade to investigate whether our focus on front-line problems, or proximate causes, means we miss the important global dynamics that drive species onto the endangered list. Essentially I will be asking whether conservationists are right to focus on poverty as the driver of the illegal wildlife trade. This requires an understanding of other threats to wildlife, and the ways that poverty and conservation interact; it also requires further investigation of the assumption that poverty drives people to poach, to encroach on protected areas for grazing and agriculture.[25] Looking at conservation from this vantage point we then have to think about whether the solutions put forward for wildlife, such as increasing anti-poaching patrols and enforcing park boundaries, are effective or not.

This raises the second major theme in this book: why do some attempts to conserve wildlife end up pitting local communities against conservationists? I will explore why clashes continue between local communities and conservation schemes around the world, and attempt to explain why they are regarded as unjust impositions, despite their good intentions. This requires us to face the difficult challenges in conservation, to delve into its darker side and examine the messy realities of apparent conservation successes. The purpose is not to make us throw our hands up and stop supporting conservation – far from it. Instead we need to examine what the real costs and benefits of conservation are. It is critical that we tackle these uncomfortable realities head on, so we can develop better conservation practice on the ground – for people and for animals.

This book will also examine the role conservation plays in producing and sustaining the 'illicit'. In one sense, the development of new conservation rules can define and sustain criminality. It is important to think about what 'criminal' means and how criminal behaviour gets defined and sustained by conservation rules. When wildlife reserves are established, local communities can suddenly find that their everyday subsistence activities have been outlawed and they have been redefined as criminals. Chapter three explores the different ways in which well-intentioned conservation activities alienate local people as they get rebranded as enemies of conservation.

As part of this I investigate how the idea of wilderness underpins much conservation practice, and how this leads to the creation and maintenance of people-free protected areas. This drives the need to engineer wilderness. Some of the world's best-known pristine wilderness areas are, in fact, engineered environments. Creating a national park means drawing up new conservation rules which outlaw the everyday subsistence activities of local communities, such as hunting for food and collecting wood.

The creation and maintenance of wilderness areas has also relied on more extreme and violent methods of control, including execution. This book will throw light on the ways that violence is deployed against local communities in the pursuit of the 'global interest' of saving species from extinction. It will investigate how this is increasingly linked to the lucrative international tourism market, which has made some species extremely valuable. The use of violence is only part of the story. One of the main themes running through all the chapters is that we need to understand how and why conservation schemes have led to much more prevalent and insidious forms of exclusion and marginalization.[26] All around the world millions of people have been, and still are being, displaced by a whole series of rules and regulations that criminalize them. One important arena for further investigation for this book is how these more invisible forms of exclusion and marginalization are dressed up in the language of partnership and participation, coupled with promises of new jobs in the tourism industry.

The first chapter sets out the profile of the legal and illegal wildlife trade, examining the scale of it and what is driving it. It then goes on to explain why enforcing the law has proved to be so difficult, looking at the role of organized crime in wildlife trafficking. Chapter two moves on to explore the world of global conservation practice. It examines the underlying ideas of conservation, biodiversity and wilderness, to explain how the removal and exclusion of local communities has been justified in defence of wildlife; the chapter points to the perils of policy making at the international level and the unjust effects on the ground. Leading on from this, the third chapter confronts one of the most controversial issues in conservation: poaching and anti-poaching. It explains how conservation ends up alienating people because of the use of force.

Chapter four explores why the ban on the rhino horn and ivory trades continue to be challenged and resisted – it is an issue that has caused bitter divides and the chapter explains why some Southern African countries are convinced that a fully legalized ivory trade might actually save elephants. The fifth chapter points to the importance of understanding the wider global context of conservation initiatives and examines the ways that mineral extraction poses a threat to wildlife. It shows how demand for coltan for mobile technology is sustaining a conflict in the Democratic Republic of Congo which threatens the last remaining gorillas. It also looks at how our love for sapphires has destroyed important habitats for lemurs in Madagascar. Finally, chapter six tackles the issue of tourism as 'the answer' for wildlife conservation, and highlights the damaging effects of souvenir hunting and the ways that landscapes are redesigned to meet tourist expectations.

The purpose of this book is to explore the ways in which wildlife conservation focuses on immediate and localized problems; it examines whether this approach prevents us from recognizing and tackling wider global dynamics that drive species onto the endangered list. The book is devoted to understanding the complex realities of conservation, and that includes exploring its darker side, the very real costs that are felt by the world's most marginalized communities. Whether we like it or not, human communities are involved in actively managing wildlife habitats around the world. It is vitally important to understand this and engage with it if we are to design and implement better conservation practice on the ground.

CHAPTER ONE

THE INTERNATIONAL WILDLIFE TRADE

ALL OVER THE WORLD wildlife is consumed as food, medicine and clothing. I have often flicked through menus and seen shark's fin soup or bird's nest soup and wondered if it contains real shark fins and real bird's nests. If the restaurant does use these ingredients then I wondered where the chef gets them from, and how these exotic delicacies were harvested. One place that bird's nests are collected is Thailand; in the limestone outcrops between Phuket and Krabi I have watched the collectors climb up precarious bamboo poles tied to the cliff sides. But I did not know what happened to the nests once they had been collected – maybe the collectors sold to a dealer who exported them all over the world. I wondered what other products were traded alongside the nests and whether all of the products were strictly legal and sustainable.

On one of my fieldwork trips to Madagascar I noticed a large ship anchored off Nosy Be island. About fifty people were walking slowly, ankle deep in sea water, and it seemed that they were staring

1.1 Collecting birds' nests, Krabi, Thailand (2008)

intently at their own feet. It made quite a startling picture. I asked islanders and representatives of local conservation organizations why so many people seemed so interested in staring at their feet in the sea, and where the ship was from and why it had come to Madagascar. Again, collecting wildlife for food was the answer. The ships came from China and local people were paid well (by local standards) to walk around the seashore at low tide looking for an expensive delicacy: sea cucumbers. There is no local market for the sea cucumber: people in Madagascar do not eat them. But conservation organizations pointed out that it was easy to get people to collect them because it was an opportunity to earn some money. The sea cucumber collection was not legal, but the ships would be in and out so fast it was impossible for the Parks Department and other government agencies to respond. I was told the same story about the Tulear area in southern Madagascar. To conservation organizations it seemed the ships were simply hoovering up the seas and taking advantage of government weaknesses.

We tend to blame poaching and illegal wildlife trading on poorer communities. We think they do it because they are greedy and do not care about precious wildlife, or because they are so poor they are forced to poach and smuggle to feed their families. But the examples of bird's nests and sea cucumbers show that the wildlife trade is driven by wealth, not poverty. As with other products such as rice and cotton, the direction of the wildlife trade is mostly from the poorer parts of the world to richer parts of the world. Some of this demand is met by the legal trade in wildlife, but a large slice of it is met by illegal trafficking.

The trade in wildlife poses a major challenge for conservation. The reality is that wildlife trade networks, legal and illegal, span the globe and are worth billions of dollars every year. This trade has spawned international agreements to monitor and control it, high-profile NGO campaigns to encourage us not to buy illegally harvested wildlife, and a variety of enforcement agencies to control it. Contrary to the often simplistic messages about stopping poachers or the trade in wildlife, the problems posed are extremely challenging. Enforcement has proved to be a very difficult business indeed. Part of the answer lies in the lucrative profits to be made through trafficking of endangered species, which attracts organized crime. This chapter will explain the shape of the legal and illegal wildlife trade and seek to explain why it still presents such a problem for conservation by examining the successes and failures in enforcement.

THE WILDLIFE TRADE

The size of even the legal international wildlife trade is staggering; worth approximately US$160 billion per year, it is one of the most valuable businesses in the world. Timber and seafood are the two largest and most important categories of trade. These figures only cover wildlife that is traded across international borders and is therefore recorded; but there are significant levels of trade in wildlife within countries which goes unrecorded. For example, turtles, fruit bats and lemurs are a common food source in Madagascar and are traded internally, so their value does not appear in international statistics. The scale of illegal trade in wildlife is almost impossible to determine precisely because it inhabits a shadowy world of illicit activities and criminal networks. But in 2007 the US State Department estimated that the global illegal trade in wildlife and plants was worth US$10 billion, but possibly as high as US$20 billion.[1]

The scale of the wildlife trade is so great that it has sparked fears that it is driving some species to extinction. Rhino horn, ivory and tiger bones are perhaps the best-known illegally traded substances, but there are many more illegally traded animals and plants that do not grab the headlines, but result in just as much damage for individual species and for the ecosystems they inhabit. Such issues have raised sufficient concern for the international community to develop a global convention to manage it: the Convention on the International Trade in Endangered Species (CITES, discussed more fully in chapter two), which determines what species can be traded, and how, and has the power to ban international trade in species if it threatens their survival in the wild. One of the earliest and most important bans agreed by CITES was on the trade in rhino horn in 1976, in response to fears that the demand for powdered rhino horn for Traditional Chinese Medicine (TCM) would lead to the extinction of the species.

In one way or another most of us are consumers of wildlife. TRAFFIC was established in 1976 as an international wildlife-trade monitoring network, governed by a committee made up of WWF and IUCN (its two partner organizations). It provides research on the legal and illegal wildlife trade, and runs campaigns to raise awareness about particular issues in the trade.[2] TRAFFIC points out that 'as human populations have grown, so has the demand for wildlife. People in developed countries have become used to a lifestyle which fuels demand for wildlife; they expect to have access to a variety of seafood, leather goods, timbers, medicinal ingredients, textiles'.[3] It is tempting to think of the illegal and unsustainable wildlife trade as a problem which afflicts poorer parts of the world – Africa, Asia, Latin America – but this is definitely not the case. The trade is driven and sustained by wealthier communities around the world. Inevitably, people who live in poverty will want to use the resources around them, and when a particular species in their area becomes internationally valuable it is only to be expected that they will engage in trading it to help make ends meet.

Wealthy industrial economies such as the UK and the USA remain major legal and illegal importers of wildlife. In 2008 a report by TRAFFIC-ASIA examined the drivers of the illegal wildlife trade in the region, and concluded that the increase in illegal trading of wildlife was directly related to the rise in incomes in the region. The report detailed the complexity of the networks involved in the wildlife trade: it linked local-level rural harvesters, professional hunters, traders, wholesalers and retailers with the final consumers of wildlife, who lived thousands of miles away from the product source. The wildlife trade also provides varying levels of economic support to different communities across the world; for some it is a regular source of income, for others it is a safety net when other sources of income dry up, and for others it is a lucrative business which generates very large profits.[4]

1.2 Collecting cicadas for food, Maerim Valley, Northern Thailand (2008)

THE GLOBAL TRADE IN MARINE LIFE

While rhinos, tigers and elephants attract the greatest degree of public attention, it is important not to forget that the largest proportion of illegal trade in wildlife is in seafood. In 2005 the British Marine Resources Assessment Group (MRAG) provided a conservative estimate that illegal fishing in Africa could be valued at approximately US$1 billion every year – a substantial economic and environmental loss for one of the poorest regions of the world. It was estimated that in Somalia the total annual value of illegal fishing for tuna and shrimp alone was worth US$94 million, while illegal fishing for sardines and mackerel in Angola was estimated at US$49 million each year. The figures are staggering. In just one illegal fishing incursion off the Tanzania coast, long-line fishing vessels from Taiwan illegally caught approximately US$20 million worth of tuna.[5] Illegal fishing generates huge profits, which means it is a clear target for organized crime and transnational trafficking networks. This is certainly the case with the international market in luxury seafoods.

We are all familiar with caviar as a luxury food product, but there are genuine fears that caviar trade and consumption could lead to extinction in the wild. The fish that produce caviar are under threat from overfishing, pollution and habitat degradation. Caviar is a perfect example of a wildlife trade which has both legal and illegal dimensions, and which is driven by wealth rather than poverty. Caviar is the eggs (or roe) of sturgeon and paddlefish, and approximately 90 per cent of globally traded caviar is produced in the countries that border the Caspian Sea. The most sought after and most expensive variety of caviar is, of course, that of the Russian Beluga sturgeon, which takes around twenty-five years to reach maturity and begin producing the eggs. The caviar is very much in demand in the world's wealthy states and TRAFFIC estimates that Beluga caviar can fetch €600 per 100g in delicatessens in Europe and the US.

One of the problems with the caviar trade is that some types of caviar (such as the Baltic sturgeon) are listed under CITES Appendix I, which means a total trade ban, while others are listed under Appendix II, which allows for strictly controlled trade.[6] Caviar is one of the few examples of a 'split listing' under CITES, where trade from some countries is banned, while it is permitted from others. This means it can often be very difficult to enforce CITES regulations because illegally fished caviar can be disguised and then traded as legally produced caviar. There are tensions between member states in CITES over regulation of the caviar trade. Sturgeon specialists continue to argue over whether stocks are recovering and whether any fishing at all is sustainable in the longer term. In the last ten years there have been calls for a complete ban on the caviar trade to allow the species time to recover; but these have been met with counter-arguments that it is an important economic sector for some areas of the Caspian Sea and a complete ban would unfairly punish legitimate and sustainable producers.

The caviar trade highlights the links between wealth and poverty and the ways in which the international wildlife trade is sometimes facilitated and driven by wider changes in world politics. Caviar production became a vitally important source of income for local communities in the Caspian Sea area following the collapse of the Soviet Union, and economic problems in Iran also prompted unemployed locals to switch to caviar fishing, some of it illegal. Despite the expansion of legal aquaculture (effectively caviar farming) in the Caspian Sea, a large slice of the international trade is met by illegal fishing. For example, in 2001 the illegal caviar catch was 12 tonnes, which was probably ten times the size of the legal catch.[7] The illegal trade meets demand in Europe and the USA, and is dependent on the use of false permits.

It is not just economic problems in the Caspian Sea area that explain the challenges of the caviar trade. The high value of the product has attracted the attention of organized crime: illegally produced caviar is much cheaper than legally fished caviar, but can still generate huge profits. In 2003 police officers in London ran Operation Ribbon, in which they raided a series of shops on Kensington High Street and uncovered a link between illegal caviar and the Russian mafia, involving extortion, violence and even murder. The police confiscated 200 tins of caviar, which were labelled as legal but were illegally smuggled Russian caviar; none of the shops were prosecuted because it was determined that they had purchased the caviar in good faith.[8]

It is safe to say that if the volume of caviar that is traded illegally is ten times that of the legal trade, then it is more than likely that any caviar on your plate has been illegally and unsustainably fished, and the profits have gone to organized crime rather than poorer local communities in the Caspian Sea. Policies aimed at shutting down illegal fishing in the Caspian Sea have to tackle this wider context and educate consumers about the problems caused by buying illegal caviar.

Caviar is not the only high-value fish product that is linked with organized crime. Another important species, and one also linked with the illegal drugs trade, is abalone, a large species of shellfish which is found in most of the world's coastlines, especially in colder waters off South Africa, Australia and New Zealand. Abalone has long been an important foodstuff for human communities, and the iridescent shell is a popular material for jewellery. The meat is considered a delicacy and is consumed mostly in East Asia (China, Korea, Japan) but also in the US, New Zealand and Chile. As with other lucrative wildlife, abalone farming has also begun to meet the demands of the global market. Abalone harvesting is usually restricted and done under licence to prevent over-exploitation, but South Africa and New Zealand present real problems for legal and sustainable fishing.

South Africa's abalone population is perhaps under the greatest threat from illegal fishing; but the drivers of illegal fishing are highly complex and stretch out through South Africa, the wider Southern African region and on to East Asia. In South Africa legal abalone fishing was done under licence until 2007; it required a government permit which was issued on an annual basis. In 2004/5 South Africa issued licences to 300 people, but the industry as a whole employed approximately 800 people. Since the early 1990s, the South African government has been concerned about the apparent increase in illegal harvesting. Records of confiscation between 1996 and 2006 show a tenfold increase in the amount of illegally caught abalone, most of which was exported to Hong Kong.[9] It is difficult to determine how much illegally caught abalone sells for, but some estimates suggest it goes for between US$100 and US$ 300 per kilo.[10] In an attempt to control the trade the government has consistently reduced the amount of catch they allow, so that in 2006/7 the total allowed catch was just 20 per cent of the catch allowed in 1995/6 (125 tonnes compared with 615 tonnes). The South African government applied to CITES for an Appendix III listing (which means the trade is closely monitored and controlled), and announced a ban on all commercial harvesting of abalone in 2007. However, illegal harvesting has continued, and estimates put the illegal catch at six to ten times greater than the legal catch. The South African government estimated that it costs just over US$3 million a year to deploy seventy officers twenty-four hours a day to patrol the abalone fishers, and this is on top of support provided by the Parks Department.[11]

TRAFFIC raised concerns that the 2007 ban would simply drive the trade underground, and encourage commercial fishermen to start working illegally. To combat this, the South African government said it would encourage the unemployed fishermen to use their skills in tourism, primarily whale watching and cage diving with great white sharks. The South African government has also promised greater investment in farming of abalone, but the problem is that it takes five to seven years for the shellfish to mature; this means it is highly capital-intensive at the start and takes a long time to generate a profit. Here again we see there are sharp differences over whether criminalizing a trade in wildlife is the best way to ensure its conservation. Indeed, one part of the explanation for the upsurge in poaching in the mid-1990s was that the local mixed-race (or 'Coloured') communities felt that fisheries should be opened up to allow a more equitable distribution of fishing rights; this built on historical grievances dating from the apartheid era when Coloured communities were able to make a living through fishing in the absence of many other opportunities.[12]

Regional dynamics are also important here. For instance, records show exports of abalone to Hong Kong from Zimbabwe and Swaziland (both are landlocked countries), and from Mozambique and Namibia. The South African government sees Namibia as a particular enforcement challenge because it does have one legal abalone aquaculture farm which exports to East Asia; this means that illegally caught abalone from South Africa could be hidden under the cover of legal farmed abalone. The fresh abalone is dried or canned in South Africa to make it ready for export; the dried abalone weighs only 10 per cent of the fresh weight, which makes it easier to hide and smuggle across international borders.

The dynamics of the illegal abalone industry are also interlinked with trades in other illegal goods. It is initially hard to imagine how abalone might be linked to a massive crime wave and a drugs epidemic, but sadly it is. The rise in value of abalone in the late 1990s attracted the interest of drug gangs in the Cape Flats area of Cape Town. Researchers at South Africa's Institute for Security Studies found that the gangs took control of large parts of the abalone fishing industry, and it was then not a huge leap for them to begin bartering abalone with other criminal networks in Cape Town. In particular, the value of abalone brought Chinese Triad gangs already active in South Africa to Cape Town, from where they could export the shellfish to Hong Kong and elsewhere. The Triads paid for the abalone with the chemicals (ephedrine and pseudoephedrine) used to make methamphetamine, a crystal form of speed which is known locally as 'tik'. As a result, in 2005–07 Cape Town experienced a 200 per cent surge in drug-related crime, and it seemed that parts of the city were gripped by a tik epidemic.[13]

The easy answer is to call for better enforcement and a crackdown on poaching. But this is too simplistic. Eliminating illegal abalone harvesting also has to involve dealing with much more complex social, economic and political problems. For instance, the smuggling routes used for abalone also carry a range of other goods including counterfeit clothes, other synthetic drugs, guns, cigarettes, diamonds and trafficked human beings. The 'abalone problem' is not something that can be dealt with in a one-dimensional way through more fisheries patrols, banning commercial fishing or the development of viable and legal abalone farms; instead it will involve dealing with the reasons why some communities in Cape Town have been gripped by tik, which opens hard questions about the legacies of the apartheid era, violence, poverty and lack of opportunity in areas like the Cape Flats. Furthermore, it would have to tackle Triads and the other transnational smuggling networks that move the shellfish from South Africa, through the Southern African region and on to Asia. TRAFFIC, among others, have argued that the 2007 ban merely pushed the trade underground, made it harder to control harvesting and alienated legitimate and licensed users. The ban may have increased the value of abalone, which encouraged Triads to get involved, driving a new drugs craze in South Africa. These complex global dynamics present a real problem for enforcement agencies. This is not confined to the lucrative trade in shellfish either, as we will see below. The same challenges exist for those keen to save the world's last remaining tigers.

THE TIGER TRADE

The dramatic decline in wild tigers is a high-profile conservation issue. The tiger is an extremely charismatic animal and an icon for the conservation movement. Tigers make regular appearances in Western popular culture: in children's stories like *The Jungle Book* (Shere Khan) and *Winnie the Pooh* (Tigger), in films, in oil company adverts (Exxon/Esso); they also feature heavily in conservation NGO campaigns and as a key tourist attraction in India. In India there is a long history of cultural attachment to tigers, and as the country cements its position as one of the world's leading economies, the demand for wild tiger watching from the growing middle class has increased.[14]

The iconic status of tigers is also linked to the fact that they are on the IUCN Red List as one of the most endangered species in the world. Despite conservation programmes such as Project Tiger in India, an international trade ban on tiger parts and support for the creation of new biological corridors to allow tigers to move between habitats, they continue to decline in the wild. TRAFFIC states that tigers are 'an iconic species in danger of extinction'.[15] IUCN estimates that of the 5,000 wild tigers left, 75 per cent are found in India, with just a handful left in China. The causes are loss of habitat and the illegal trade in tiger parts. IUCN figures show that tigers have lost approximately 93 per cent of their historical range and habitat. It is estimated that tiger habitat has contracted by 41 per cent since the mid-1990s (with the greatest losses in India), and remaining habitats are increasingly fragmented. In the past, tigers were found from Turkey to the eastern coast of Russia, but many of the sub-species that inhabited these historical ranges are now extinct. Since the start of the twentieth century tigers are no longer found in south-west and central Asia, in Java and Bali; nor are they found in large areas of South-east and Eastern Asia. These days tigers are found in thirteen range states: Bangladesh, Bhutan, Cambodia, China, India, Indonesia, Laos, Malaysia, Myanmar, Nepal, Russia, Thailand and Vietnam; they may also still be found in North Korea but there is little up-to-date information on populations there. IUCN identifies the main causes of habitat loss as conversion of forest land to agriculture and silviculture, commercial logging, and human settlement. Of course the loss of habitat also means that tigers are more likely to come into conflict with local communities who may kill them to protect livestock.[16]

The second threat to tiger populations is the trade in tiger parts, which is illegal under international regulations as agreed by CITES. In the mid-1990s, when the Environmental Investigation Agency (EIA) was campaigning against the rhino-horn trade, their investigators in China and Taiwan, among other countries, found that they were also stumbling across illegally traded tiger parts. Tiger bone has long been used as an anti-inflammatory in Traditional Chinese Medicine, and other parts such as claws and fat are also used in medicines; in contrast, tiger skins are mainly used for ornamental purposes and as fur trim in clothing.

China banned domestic trade in tiger products in 1993, instituting a public-education campaign and promoting substitutes to tiger-based medicines; this approach was combined with more legalistic and punitive measures such as greater enforcement and punishment for law breakers. Recent TRAFFIC research indicates that the CITES regulations and the Chinese ban has enormously reduced the sale of tiger products in China. Undercover surveys found very little tiger bone available in China, and fewer than 3 per cent of 663 shops surveyed claimed to sell tiger products.[17]

However, the illegal sale of tiger parts does continue, albeit on a smaller scale, not only in China but in several other Asian countries, including Malaysia and Vietnam.[18] For example, an EIA undercover investigation in 2008 found fourteen shops selling big cat skins in contravention of national legislation and CITES regulations. Traders noted a decline in demand for skins from the Tibetan community following the Dalai Lama's call to end the use of big cat skins in traditional robes (discussed further in chapter 3); however, snow leopard and tiger skins were still trafficked from a diverse range of countries including Vietnam, Afghanistan, India and Russia. The estimated prices for a tiger skin ranged from US$15,000 to US$22,000 over the period 2005–8.[19] IUCN also points out that many medicines contain fake tiger bone, and that the labelling and advertising of these fake medicines sustains the wider demand for tiger products. The EIA is sufficiently concerned about the situation to make the illegal trade in tiger parts one of their major campaigns, centred on 'Species in Peril'.[20]

The problem is often attributed to a lack of political will and poor enforcement. The trade is extremely complex and stretches across a number of different countries that produce and consume tiger products. It is not just China and its neighbours that consume tiger parts: demand for TCM comes from wealthier communities all over the world. It is also clear that illegal traders and smugglers can only operate with some degree of complicity from officials, either through turning a blind eye to illegal activities or by actively assisting it.

NGOs often lay the blame at China's door for failing to do enough to prevent the trade in tiger parts, but other countries also lack proper enforcement measures. A recent TRAFFIC report, *Paper Tigers?*, states that while there is no evidence that the US captive tiger population plays a major role in the illegal tiger trade, the relevant US legislation is poorly enforced and contains a series of critical flaws. It is estimated that the US has around 5,000 tigers in captivity (in zoos, circuses and private ownership). However, the TRAFFIC report found that the US government has no way of knowing how many tigers there are in captivity, where they are, who owns them or what happens to their body parts when they die. As a CITES signatory the US has a duty to ensure that captive tigers do not enter the illegal trade in tiger parts There is a continuing illegal trade in tiger parts in the US, with products imported from East Asia to the US. For example, official trade records show that between 2001 and 2006 the US Fish and Wildlife Service seized more than 250 shipments of products, mostly traditional Asian medicines labelled as containing tiger.[21]

The persistence of trade in tiger parts has led some to suggest that a legal, controlled and sustainable trade might be the best way to conserve the species in the long term. This argument has caused deep divisions and sparked an ongoing controversy. The continuing illegal trade in tiger parts has inevitably led to claims that legalization of the trade is the best way to secure the tiger's future. The issue of the tiger trade is complicated by the existence of legal tiger farms in China. The owners of the farms have been pressuring the Chinese government to reopen a legal trade in tiger parts, claiming that the tigers in the farms are reared and slaughtered humanely. The idea behind the tiger farms is similar to that behind the creation of bear farms in China in the early 1990s: that they can service a legal trade and prevent poaching in the wild. It is estimated that there are currently 4,000 tigers in the farms, with claims that they could breed up to 100,000 in the next ten to fifteen years. This is a potentially lucrative industry; it is possible that a whole animal could fetch US$40,000 while individual parts might attract even higher prices, from US$20 for claws to US$20,000 for a skin.[22] However, the tigers have no value in the legal wildlife trade. Unlike those who own the bear farms, the owners of the tiger farms cannot legally trade their produce; instead they have stockpiled tiger parts in the hope of future legalization.

NGOs like the International Fund for Animal Welfare (IFAW) and Save The Tiger Fund are opposed to the development of tiger farms. In 2007 the fourteenth conference of CITES also agreed to call for an end to legal tiger farms in China. Opponents point to the inhumane conditions of the farms and claim that the owners of the farms are already engaged in the illicit trade in tiger parts, and therefore do not have the interests of tigers at heart. They also point out that the bear farms have not prevented bear poaching to feed the demand of the TCM trade. For numerous tiger conservation organizations, the only answer is to campaign against the use of tiger parts for TCM and definitely not to meet the demands of the trade through legalized farming.[23] In any case it would seem that any legalized trade in farmed tigers would start to reopen a market for tiger products that has already seen a sharp decline in China, effectively overturning the efforts and developments of the last fifteen years. I will return to the issue of legalization rather than banning as a solution in chapter four because it is a critically important and highly divisive issue in debates about the ivory and rhino horn trades. One of the main arguments against legalization is the difficulty of enforcing regulations so that the trade is sustainable and controlled. These failures are the result of multiple and complex challenges which are overlooked in favour of simplistic arguments and easy fixes.

ENFORCEMENT

The problems with enforcement are partly the result of under-staffing and a lack of trained wildlife officers. In the UK there are approximately 600 trained wildlife officers to cover all entry ports to the UK. HM Revenue and Customs is responsible for enforcing the international rules that regulate the trade in species of flora and fauna listed under CITES. But 30,000 species of plants and animals are listed under CITES; some are covered by a trade ban, but regulated trade is permitted for many species on the list. This makes it difficult for individual customs officers to determine whether an animal or its derivatives are being traded legally or not.[24] Although most regional police forces in the UK have a dedicated wildlife officer, some areas are still left without police expertise to deal with wildlife crime. TRAFFIC-Asia tried to tackle this by producing a booklet of species that are illegally traded in their region. The description of freshwater turtles provides a window on the problems customs officers face: the turtle is often traded in paste form, so wildlife officials and customs officers would find it almost impossible to determine whether the paste they had intercepted was from an illegally traded endangered species.[25]

It is difficult to determine which agency should be responsible for tackling organized crime networks when they engage in wildlife trading. This is reflected in the range of agencies responsible for combating wildlife crime. If a consignment of ivory is trafficked into the UK via Heathrow Airport, it is unclear whether this is the responsibility of the Department for Environment, Food and Rural Affairs (DEFRA) as the government department responsible for environmental issues, or maybe the airport authorities, or Customs and Excise, or the local police force, the national police force in the arrival country or the country of origin, or

1.3 Pufferfish for sale to tourists, Phuket (2008)

Interpol. In effect an illegal consignment of ivory at Heathrow is the responsibility of all those agencies, but for many of them the main concern will be drugs or terrorism, so it is easy for wildlife crime to slip past unnoticed.

In the UK each police force has its own wildlife officer, but the police authority which deals with wildlife crime is the National Wildlife Crime Unit (NWCU). It is based in North Berwick (in the south-east of Scotland) and draws together a range of agencies including the police, the Home Office and HM Customs and Revenue.[26] The UK government has also set up Partnership for Action Against Wildlife Crime (PAW), a sub-section of DEFRA, to co-ordinate and support the UK networks of police wildlife liaison officers and customs wildlife and endangered species officers. PAW also aims to draw attention to the growing problem of wildlife crime and to raise awareness of the need for tough enforcement action. Most police forces now have at least one wildlife liaison officer, though many of them carry out these duties in addition to their other policing responsibilities.[27] Although PAW provides co-ordination, wildlife crime can be encountered by multiple agencies including national and regional police forces, customs officers and animal groups such as the Royal Society for the Prevention of Cruelty to Animals (RSPCA). This makes enforcement very difficult.

The case of Operation Charm in London clearly illustrates the problems involved. The Metropolitan Police, responsible for the London area, started Operation Charm in 1995, and it remains the only police initiative against the trade in endangered species in the UK. Operation Charm has had some success in combating wildlife crime. Between 1995 and 2007 Operation Charm seized over 30,000 items made from endangered species. These ranged from well-known illegal goods such as leopard fur coats, stuffed tiger cubs and ivory carvings, to less familiar items such as Arowana fish from Asia, prized by aquarium enthusiasts and thought to bring good luck to owners. However, it was not until 2006 that Operation Charm had its first 'enforcement success', when the owner of a Traditional Chinese Medicine shop pleaded guilty to eighteen charges under the Control of Trade in Endangered Species (Enforcement) Regulations. He had been selling materials made from orchids, saiga antelope, musk deer, bears and seahorses.[28]

Operation Charm has mostly focused on seizures of illegal goods, and it operates via a combination of law enforcement and awareness raising through publicity campaigns. It covers TCM, fashion and decorative items, food (mainly bushmeat and caviar), and the trade in live exotic animals. The initiative has been supported by the Federation of Traditional Chinese Medicine and the London Chinatown Chinese Association.[29] It has received endorsements from international celebrities, such as the actors Harrison Ford and Ralph Fiennes, as well as the champion Chinese snooker player Ding Junhui.[30]

In 2006 Operation Charm became a partnership between the Metropolitan Police Wildlife Crime Unit, the Greater London Authority, the Active Conservation Awareness Programme (ACAP), IFAW, WWF-UK and the David Shepherd Wildlife Foundation (DSWF). This range of actors reflects the variety of organizations involved in combating wildlife crime. It also provides part of the explanation for why the wildlife trade is so attractive to organized crime. It is an extremely lucrative industry, but traffickers know there is a relatively low chance of being caught because so many different agencies are 'responsible', and the main law enforcement agencies do not have the funding, staff or support to make it their top priority.

The complicated range of organizations involved at the national level is compounded when we look at the relationships between national agencies and international agencies involved in dealing with wildlife crime. The UK response also has to integrate with international initiatives to combat wildlife crime. For example, PAW works with Interpol, an organization which deals with organized crime, terrorism, drug smuggling and people trafficking. Interpol has been dealing with environmental crime since 1992. In 2002 Interpol set up a separate section for wildlife, which is an indicator of the value of the trade and the involvement of global organized crime networks. Since 2006 Interpol has had a full-time officer to manage the wildlife crime programme; however, it has been reliant on external assistance. Interpol states that the expansion of the wildlife crime programme has been supported and assisted by conservation and animal welfare NGOs: IFAW, Safari Club International and the Bosack and Kruger Charitable Foundation. The Wildlife Working Group in Interpol has a number of duties which include maintaining an international network for the exchange of information on wildlife crime; enhancing domestic operations in each country through cooperation and coordination; assisting in the training of wildlife enforcement officers in developing countries; encouraging the integration of wildlife enforcement activities within broader international wildlife conservation initiatives; and improving international communication with member countries by coordinating and leading international meetings with regional representatives.[31] The involvement of Interpol draws our attention to the fact that global networks of organized crime are heavily involved in the wildlife trade.

One of the problems that conservationists face is that much of the illegal trade in flora and fauna is carried out by networks of organized crime. The trade is worth billions of dollars each year. The illegal trade in wildlife is so lucrative that it has attracted the attention of organized crime, including the Mafia, Yakuza and Triads. The trading networks and groups involved are often also engaged in other forms of illegal trade such as drug smuggling, people trafficking and illicit arms trading. The illegal trade in wildlife is attractive because the rewards are high and the chances of being caught are comparatively low.

Traffickers make use of new technologies to trade wildlife, and the internet has proved to be an important means of selling wildlife across the world. In 2008 IFAW published *Killing with Keystrokes*, the results of its fourth investigation of online trade in CITES Appendix I and II listed species. It followed 7,122 online auctions in eleven countries over a six-week period. Their study revealed a very high volume of illegal trade in endangered species, worth US$450,000 in total. IFAW identified enforcement as a critical problem. For example it is almost impossible to determine if the ivory offered on the internet is legal or not, and even after eBay banned all cross-border trade in ivory on its site, it was still found to be responsible for 83 per cent of ivory sales and 63 per cent of all wildlife sales on the sites being monitored.[32] Ivory was the biggest category, followed by exotic birds such as hyacinth macaws and Moluccan cockatoos. Traders are able to exploit loopholes on the internet, such as sites that allow personal ads and message boards where potential traders can make contact and then conduct the transaction privately, thereby flouting the regulations of the internet site. They can also circumvent regulations by using creative spelling such as *ivorie* or *iv*ry*, as well as using counterfeit permits and documents which are impossible to verify online.

The report also drew attention to some surprising results, which served to underline the argument that the problem in the wildlife trade is demand from the rich world. For example, the US was responsible for more than two-thirds of the trade, 70 per cent of the international trade in protected species, which is nearly ten times the value of the next two culprits, the UK and China. The internet offers a new range of challenges to enforcement agencies that are already overstretched and underfunded. Taking the example of eBay, the company has worked with IFAW to stamp out illegal trade of animals on its site; however, the policies are often poorly enforced, relying on voluntary compliance or voluntary reporting by other users. eBay Australia has an outright ban on all ivory sales, but even so a small number of ivory sales were still conducted there.[33] The problems are magnified when we consider that organized crime is not just involved in trafficking wildlife – it is also responsible for the production of illegally traded goods.

COCKLING IN MORECAMBE BAY

The case of cockling in Morecambe Bay, a stretch of coastline in Lancashire in the north-west of England, illustrates why organized crime is interested in the wildlife trade, how conservation and crime are interlinked and how the involvement of multiple agencies acts as a barrier to enforcement.

In 2003–4 something new was happening in Morecambe Bay, which is just one mile from my home. It has been worked by a small number of fishing boats and cocklers for years. Historically, cockle picking in Morecambe Bay was a small-scale local enterprise which provided seasonal employment for a few local families with permits to fish the cockle beds. The practice of cockle picking is potentially dangerous. Local residents are only too aware of how treacherous the sands can be: the tide rises rapidly, and once it starts coming in anyone on the sand has thirty minutes or less to get back to land. People are easily trapped two or three miles out in the middle of the bay, risking drowing or being caught in quicksands. Cross-bay walks are popular summertime activities, but are organized and led by the 'Queen's Guide to the Sands', who knows the channels, tides and areas of quicksand.

1.4 Morecambe Bay, UK (2010)

In 2003 this all changed. The sands were teeming with people fishing for cockles – their four-wheel drive vehicles were buzzing around and huge refrigerator lorries waited at the shoreline to load up the catch. The value of cockles spiralled from approximately £200 per ton to £1,300 per ton. The shellfish bonanza beneath Morecambe sands became a major attraction for organized crime, with up to £8 million worth of cockles in the sands at any one time.

Some of those involved in cockling in Morecambe Bay explained the rise in value as the result of a shipping accident in Spain. When the *Prestige* tanker sank off the coast of Galicia in Spain, it spilled 64,000 metric tonnes of oil, much of which polluted local shellfish habitats. The resulting closure of cockle beds in Spain prompted Spanish shellfish companies to switch to Britain to meet demand. This created a gold-rush mentality. The cockle bonanza was so lucrative that it started to attract organized criminal networks that specialized in the provision of cheap labour through people trafficking, primarily from China. This produced stiff competition, and even violence on the sands, between rival groups of cocklers who wanted to maximize their share of the profits. Some cockle pickers even needed 'minders' to ensure their safety on the sands.

The bay is a public fishery, and so it is hard to regulate who uses it.[34] The North-West and North Wales Sea Fisheries Committee issues the permits for cockle fishing in the area. These licences are intended to provide some level of protection for the cockle beds through the prevention of overfishing, and stipulations about removal of undersized cockles. The sands are extensive (120 square miles according to the coastguard services) and with little police presence in the area it was easy for the illegal cocklers to avoid detection. As an individual involved in prosecuting the case put it: cocklers are up to three miles out on the sands, and the Department of Works and Pensions do not have the staff to send people out to check whether they have work permits or not.[35] This made illegal cockling attractive: large profits could be made with little chance of detection.

Until 2004 the story of the arrival of criminal gangs, illegal labour and the cockle gold rush did not really register in the public mind. It was the deaths of twenty-three cockle pickers that shed light on this industry. Late on the freezing cold night of 5 February 2004 emergency services were called out for a search and rescue mission for a group of Chinese cockle pickers stranded on Morecambe Bay. Their gangmaster, Lin Liang Ren, had transported the cockle pickers to the sands, provided them with housing and supplied them with tools. They were out on the sands late at night in order to fill two lorryloads of cockles.[36] Lin had accepted the order even though it meant his cocklers would

1.5 Abandoned vehicle used to transport cockles to the shoreline (2010)

have to work through the night in dangerous conditions to meet the deadline. The Chinese cocklers did not know about the hidden dangers of the sands until it was too late, and Lin delayed calling the emergency services for fifty minutes after it was clear that the cocklers were caught by rising tides. Fearing the worst, many of the cocklers made calls home to relatives in China to say goodbye. Some of the survivors of the disaster were so fearful of being found by the authorities that they tried to hide from the coastguard rescue helicopters.

The cockle pickers themselves had been trafficked into the UK from Fujian by Chinese Snakehead gangs who had charged their families up to £20,000 for passage to the UK. Many of the pickers were locked into the industry, afraid to escape or speak out about conditions for fear of deportation. The local member of parliament, Geraldine Smith, pointed out that one of the problems was that the Chinese government refused to recognize illegal Chinese immigrants as Chinese citizens; this meant that if police and immigration services detained the workers, they could not be deported back to China because the Chinese government would not accept them.[37] In effect they would become stateless people, so there was little point in the authorities in the UK detaining illegal Chinese workers.

The pickers were paid a mere £5 per sack of cockles, while Lin would have received three times that amount, and the company that he sold them to would have made an even greater profit. In 2006, after a protracted investigation and legal proceedings, Lin was convicted of the manslaughter of all twenty-three cockle pickers. The company that bought cockles from Lin was cleared of any wrongdoing; it was determined that the companies could not have known that the use of Chinese labour would have involved illegality.[38] The newspaper stories and court case revealed glimpses of the terrible human stories behind the incident. For instance the families of the dead workers still owe the debt to the gangs that trafficked their relatives to the UK; the workers had hoped that a short period of work in the UK would provide their families with a better life, and had fully intended to return home after a few years. Some of the dead were parents who had left their children with relatives, so that they could work to give their families a better life.

The deaths seemed to come out of nowhere, but in fact UK government authorities had been made aware of the potential for such a disaster prior to 2004. Interestingly the first alarm bells were sounded not from the sands, from conservation agencies or from immigration authorities, but from local traffic police working on the M6 motorway, who had been stopping the vans used to transport the workers from Liverpool to Morecambe Bay because they were not roadworthy. Local MP Geraldine Smith had written to the Home Office in June 2003 to express concern about the safety of the Chinese cocklers. She complained that they were being transported in overloaded boats across dangerous waters and that they were being paid a fraction of the usual rate for cockle picking. She had also tried to get the Department of Works and Pensions to see the bay as a workplace, and invoke Health and Safety regulations. But because it was a coastline it was difficult for officials to see it in the same way as a factory or an office.[39] At the time, Fiona Mactaggart, the Home Office Minister, said the immigration service had too few resources to deal with the problem. An investigation was ruled out, because it would 'serve little useful purpose'.[40] Following the trial in 2006, the UK government was criticized for not heeding these earlier warnings.

The whole episode has had a serious impact on the environment in the Bay. In fact, in 2005 the beds were closed to legal cockling by the North West and North Wales Sea Fisheries

1.6 Notice announcing that the cockling beds are closed (2009)

Committee because they had been overused and the
failed to reproduce. These days it is mussels that attract
workers and refrigerated lorries from Europe.[41] The g--
demand for mussels, and seed mussels for Dutch mussel farms,
has reshaped and reorganized shellfish collection in Morecambe
Bay but the underlying dynamics remain the same: consumer
demand in wealthy countries for wildlife drives the involvement
of global criminal networks in places all around the world.

CONCLUSION

While big charismatic species like rhinos and tigers might grab
the headlines, the illegal wildlife trade is massive, and tackling it is
extremely challenging. The wildlife trade is driven by wealthy
states, such as the US and UK, as well as by affluent communities in emerging economies, like China and India. People all over
the world consume wildlife in a variety of ways: as medicines,
ornaments, foodstuffs and clothing. The trade in shellfish demonstrates very clearly that those involved in the illegal wildlife trade
are also engaged in other criminal activities; in this sense they are
not 'wildlife specialists'; instead, trafficking lucrative endangered
species is one part of a wider portfolio of illicit activities. Cockling
in Morecambe Bay is a good example of this. The trade in
shellfish is legal, but the large profits mean it is attractive to
organized crime; these networks then provide a cheap illegal
labour force to extract shellfish without a licence, which allows
crime networks to capture maximum profits from the trade. This
presents even greater enforcement problems than the illegal trade
has attracted in organized crime; wildlife can be moved along
the same international trading routes used for drugs, people
trafficking and arms smuggling. Wildlife enforcement officers are
well aware of this, and are keen to point out that when someone

is convicted of wildlife crime, it is highly unlikely that they will not already be known to law enforcement agencies. Therefore, standard one-dimensional approaches such as increasing patrols in national parks or fisheries, or stepping up firepower for park rangers, are not going to work in the longer term. Taking the case of caviar, it is linked to changes in world politics such as the break-up of the Soviet Union, and to the need for waged employment in poorer areas around the Caspian Sea, but most critically it is driven by wealthy consumers keen to eat a luxury product. The same can be said for abalone, a trade which is linked to drug production and consumption, but which ultimately services a market for affluent communities in the Far East and wealthy consumers in the West.

The complex dynamics that drive and sustain the wildlife trade mean that it is very difficult to enforce regulations designed to conserve species. Conservation and crime are linked in two different ways. The first way is via the illegal trafficking of wildlife and involvement of organized crime. The second link is the ways conservation rules themselves define and sustain crimi-nality, and then create friction and resistance, which is the subject of the next chapter. Chapter two examines why well-meaning conservation initiatives often end up pitting local communities against wildlife and reveals the perils of policy making.

CHAPTER TWO

GLOBAL ACTION, LOCAL COSTS

SEAHORSES ARE A FAVOURITE sea creature, but the international trade in dried and live seahorses is endangering them in the wild. Some 90 per cent of trade in seahorse specimens is for Traditional Chinese Medicine; they are also sold dried as curios and as live animals for commercial aquariums or as pets. Under CITES this trade is permitted as long as the producer country can prove that trade in seahorses is sustainable and does not endanger the animal in the wild. But the seahorse trade draws in all the countries that produce seahorses, including Vietnam, Australia, Mexico, Indonesia and Sri Lanka, as well as all the countries that consume them (China, USA and many members of the European Union).[1] It stands to reason that there is little point in Vietnam or Mexico attempting to regulate the trade on their own. Instead it is more effective to work towards a system of international co-operation to ensure that the trade in seahorses does not lead to extinction. Consequently, wildlife and the international trade in it is a major focus for international cooperation. Wildlife populations cross international boundaries, which leads to calls for greater levels of global co-operation. This is especially important for species that migrate long distances, such as birds, butterflies, sharks and whales; but it is also critical for wildlife populations which cross national borders as part of their everyday search for food, like elephants and mountain gorillas. There is no point in one country putting all its efforts into conservation if the neighbouring country does little to protect wildlife.

International regulations define what we see as legal and appropriate use of wildlife versus what we see as illegal and inappropriate use. This shapes conservation practice at the national level and is crucial to understanding how and why conservation works in some places and not in others. In essence, international regulations revolve around standardization: a global 'one size fits all'. This chapter deals with the second question of this book: how is it that well-intentioned initiatives so often pit local communities against conservation?

THE CONVENTION ON THE INTERNATIONAL TRADE IN ENDANGERED SPECIES (CITES)

Fears about the impact of the wildlife trade on species conservation led to the development of CITES in 1973, and it came into force in 1975. Its primary objective is to ensure that international trade in wild animals and plants does not threaten their survival. Membership is voluntary, and is made up of national governments; in 2008 there were 173 member countries. The CITES Secretariat is based in Geneva, Switzerland, and plays a key role in disseminating information to member states, organizing the biennial Conference of the Parties (CoP, the main decision-making body) and acting as an advisory body for the Convention. CITES operates through a designated National Management Authority in each member state, and each country designates one or more 'Scientific Authority' to advise them on the ways that trade might be affecting species.[2] Although CITES is an international level convention, it is still heavily reliant on national-level enforcement and monitoring.

CITES uses a system of Appendices as its main management 'tool'. If a species is listed under Appendix I – as are elephants, rhinos, leopards and tigers – then trade in it is effectively banned. Species listed under Appendix II are endangered but not necessarily threatened with extinction; therefore, trading them is permitted but it strictly regulated to prevent overuse. Examples include the hippo, the vicuna and the saiga antelope (whose wool is highly valued in the fashion industry, to make pashminas). An Appendix III listing covers species that are protected in at least one country, and where one of the member states has asked fellow members to assist them with regulating the international trade in that species; for example, blackbuck antelope and golden jackal are listed under Appendix III, because India and Nepal have requested assistance with conservation and regulation of any international trade.[3] It is possible for species to move between Appendixes, but this requires a two-thirds majority agreement at the biennial conference. In rare cases populations in some countries are listed under Appendix I (effectively a total ban on international trade) while other countries have lobbied for an Appendix II listing (restricted trade). African elephants fall into this category, and the controversies surrounding this 'split listing' are covered in chapter four.

Members of CITES have a common goal to prevent trade threatening species with extinction, so they sign up to a series of global agreements to harmonize species conservation across international boundaries. However, getting and maintaining a consensus on these international agreements has proved to be very difficult. The first problem is that the main management tool of CITES, the appendix listings, are presented as scientifically determined and therefore politically neutral.[4] Scientific knowledge is, naturally, important in global approaches to conservation, and wildlife management has become a specialized and technical arena. There is therefore a tendency to assume that scientists are politically neutral, and global conventions rely on claims that the scientific conclusions on which they are based is not only uncontested but can provide neat and unproblematic answers.

This means that the scientists working for international conventions have a very powerful position: they are able to define the 'problem' for conservation, and then suggest technical solutions to those problems. They decide what can and cannot be done with wildlife, and what is defined as legal and illegal. Scientists and experts thus act as powerful 'knowledge brokers' (or epistemic communities) – they inform, guide, and even determine policy outcomes at the local, national and global levels.[5] The problem is that they are usually not able to deal with variations at a local, regional or national level; nor can they take account of social and cultural issues which may arise across countries and regions.

In most cases species are listed under one appendix (I, II or III), which treats the species as a single global population. This means that variations at the national level are ignored; in one country a particular species may not be endangered, but in the neighbouring state it is. Countries with healthy populations face very different management challenges than those with dwindling and threatened populations. To make matters worse, member states who want to use and trade their healthy wildlife populations have to prove to CITES that any trade will not have a negative impact on the species. This is referred to as a 'non-detriment finding'. Members of CITES who want to trade often feel irritated that they are required to produce a lot of evidence to prove that trade will not endanger wildlife, whereas members in favour of a ban are not asked to do so. This is because CITES policy, like many environmental laws, is based on the idea of the 'precautionary principle': that because we cannot necessarily know whether trade is damaging wildlife populations, we must err on the side of caution. While this is presented as scientific and politically neutral, it is not, and it produces fierce political debates within CITES.[6]

Despite the claims that international legal frameworks draw on scientific information and produce neutral management instruments to save species, it is clear that any decision is highly political. Maintaining consensus on global agreements within CITES is also difficult because of the relative powers and roles of states and NGOs. Although only the representatives of national governments can actually vote on CITES decisions, NGOs such as WWF took a major part in lobbying for the creation of CITES, and they now play an important role providing information, lobbying and public campaigning at the international and national levels. NGOs have been central in arguing for trade bans as a moral issue to save certain species, such as elephants, rhinos and tigers;[7] they are brilliant at bringing issues to the attention of CITES members and providing information that can be used to shape and determine whether a trade is banned or not. This is very much the case with the ivory ban: NGOs like the African Wildlife Foundation and WWF determined the acceptable limits of the debate and effectively pressured members to vote for a total ban (this is discussed more fully in chapter four). At times the positions taken by NGOs have resulted in some fractious disagreements between such organizations and member states.

The 'one size fits all' policies pursued by CITES are, inevitably, open to accusations that leading members of the convention get to determine trade policy. This is especially fraught when dominant actors decide on policy towards wildlife they do not have to manage within their own countries. So for example, all members of CITES get to vote on trade policy over tiger skins, ivory and rhino horn. Since so many of CITES' decisions restrict what can and cannot be done with wildlife in developing countries, CITES is often criticized for having a 'Western bias'; there are regular complaints that vocal and dominant members from Europe and North America get to decide how African states can use their elephants or Asian states their pangolins. If a species is listed under Appendix I then trade is banned at a single stroke, and is based on the assumption that a trade ban will give populations time to recover. This kind of blanket ban cannot take local or national differences in wildlife populations into account, and so it produces serious tensions within the Convention. These decisions are not likely to solve conservation problems in the long term. For critics, CITES has drifted from its original purpose of managing and regulating trade, and is now a convention that focuses on banning trade and issuing declarations about the need for enforcement of bans. The dominance of wealthier Western states has caused resentments, and many developing countries view CITES as another international institution under whose auspices their wishes and best interests have been overridden by those of the West.[8]

One of the major challenges faced by CITES is the implementation and enforcement of its decisions. It is one thing to decide to ban the trade in tiger bone or to monitor and restrict the trade in tortoises, but without implementation on the ground these decisions are nothing more than symbolic gestures. CITES is poorly enforced across the globe. There is a lack of adequate national legislation, and members often fail to provide timely and accurate reports on seizures and infractions. It would be too easy to argue that the record of the developing world is worse than the record of wealthy countries, but that is simply not the case. The US has some of the best national legislation, in the form of the Lacey Act and the US Endangered Species Act,[9] which cover the trade in endangered species but it is poorly enforced. The fact is that the wildlife trade is simply not a high priority for most countries which have other competing demands on their budgets. The demand for wildlife products in the wealthier communities of the world completely overwhelms the existing enforcement structure.

The problems with enforcement are also a result of the way the CITES system is organized. CITES allows and monitors trade through a permit system; in theory this is fine, but permits can be falsified and genuine permits can be found on the black market. For example, the six-month-long presidential crisis in Madagascar in 2002–3 produced chaos in government ministries. During the crisis hundreds of blank CITES permits were stolen from government offices and were used to illegally export reptiles and amphibians (mostly frogs, tortoises and geckos) for the international trade in pets and ornamental creatures.[10] So the permit system is only as good as the people and system responsible for overseeing it at the national level. Reports by CITES and NGOs like TRAFFIC have pointed out that permits are regularly forged, while genuine permits could easily be bought from officials.

Conventions like CITES also face challenges because they are founded on fuzzy concepts like biodiversity and wilderness. These are culturally created ideas, and lack clear definitions. Yet they are the founding principles of international conventions and are backed by increasingly powerful global NGOs. In order to understand why global conservation is resisted at the local level, it is useful to examine one of the core ideas of conservation practice: biodiversity.

BIODIVERSITY

Biodiversity is an important idea which underpins conservation thinking, and has become increasingly influential in global conservation practice. The Harvard University scientist E. O. Wilson is credited with bringing the term into common usage, via his edited collection *BioDiversity* (1988), the proceedings of the US National Research Council meeting on Biological Diversity. In 1992 he published *Diversity of Life*, which set out the importance of conserving ecosystems.[11] Wilson's idea of biodiversity has acquired global authority, being used as the founding principle for the Convention on Biological Diversity (CBD), which came into force in 1993, following a series of meetings and negotiations in the late 1980s.[12] Its purpose is conservation of biological diversity, sustainable use and equitable sharing of genetic resources.

As part of Wilson's life's work he set up the E. O. Wilson Biodiversity Foundation, which continues to inform and shape debates about what biodiversity means and why it is important to conserve it. The Foundation is organized around a 'triple string' of science, education and business; to quote its mission statement: 'The mission of the E. O. Wilson Biodiversity Foundation is to preserve biological diversity in the living environment by inventing and implementing business and educational strategies in the service of conservation.'[13] (This phraseology is indicative of the way in which conservation is increasingly promoted in terms that are 'business-friendly'; in fact as part of its fund-raising activities the Foundation offers experiences and opportunities for sale on eBay, including allowing potential customers to bid for lunch with Harrison Ford, or to name a newly discovered species of orchid.)

Everyone thinks they know what biodiversity means. It is related to the diversity of life, and indicates the existence of a wide variety of species in a particular place. In fact it is far from having a single agreed definition. For instance, the CBD defines biodiversity as the variety of species, but also includes the genetic differences within species, because they are defined as the building blocks of life which determine the uniqueness of each species and each individual within a species. Interestingly, the Convention also defines biodiversity as 'the web of life' which sustains the planet, including human life.[14] Rather like other environmental shorthands, such as global warming, the idea of biodiversity has developed during the last twenty years and been invested with varied and powerful meanings. One of the critical concepts in conservation, it is often presented as neutral, technical and scientific, and therefore incontestable.[15]

The power of biodiversity as an idea has been enhanced by enthusiastic backing from international conservation NGOs like Conservation International, WWF and The Nature Conservancy, and also by the United Nations, which has declared 22 May as International Biological Diversity Day and 2010 as the International Year of Biodiversity. Its importance as an organizing and mobilizing principle in conservation is clear, and the term is frequently used in debates and campaigns about the need to stamp out illegal trade in wildlife and in demands to create new protected areas.

However, the implications behind the concept of biodiversity have had consequences that are potentially problematic for some of the poorest parts of the world. As Bill Adams, Professor of Conservation and Development at the University of Cambridge, argues, there is an assumption that biodiversity means somewhere where there are no people, and therefore that it is found in wilderness areas that are separated from human interferences.[16] Biodiversity thus provides a rationale for the creation and maintenance of areas where wildlife is protected, but people are forced out.

2.1 Clear demarcation of park boundary and agricultural land, Bwindi Impenetrable National Park, Uganda (2007)

THE POWER OF WILDERNESS: THE CREATION OF CRIMINALS

Tourists visiting Kruger National Park, in South Africa, may believe they are experiencing a pristine wilderness, where lions, elephants and many other rare and beautiful species roam free in a timeless, pristine and people-free wilderness.

In fact, however, Kruger is a manufactured Eden. The people have been removed and the area's social and political history has been airbrushed out of the glossy tourist brochures. A closer look at Kruger reveals glimpses into the violence and forced evictions carried out to create one of the best-known 'wilderness' areas in the world. The main tourist camp in Kruger is called Skukuza; it is named after the Tsonga term for James Stevenson-Hamilton, the first warden of the park. Literally translated it means 'he who sweeps clean', a reference to his removal of all resident communities to create a space devoid of human activity.[17] The communities that once lived in the area suddenly found their everyday practices of using the area for food, shelter and fuel were outlawed: they had to move outside the new park boundaries or face being branded as criminal trespassers. These evictions were not necessarily part of a long-past history; in some areas the violent evictions are very much in living memory. In 1969 South Africa was still under the apartheid system, and in that year the Makuleke community were forcibly removed from their land and relocated outside the Punda Maria Gate of the Kruger National Park.

The creation of the Kruger National Park demonstrates one of the main themes of this book: that conservation rules can result in serious social injustices, through the ways they redefine everyday practices as criminal behaviour. Once the park was established, people who continued to collect firewood or hunt for small game were defined and treated as criminal trespassers. The creation of Kruger National Park and the violence associated with it highlights

2.2 Elephant watching in the Kruger National Park 'wilderness' (2006)

the ways in which the ideas of wilderness and biodiversity are used to mobilize public support for setting aside land for conservation and for removal of communities to create wilderness areas.

This vision of wilderness is often seen to derive from Yellowstone National Park in the USA, which was formally created in 1872. As Jacoby's study of the origins of Yellowstone, the Adirondacks and the Grand Canyon national parks shows, local communities were dispossessed of their lands and prevented from using resources in the newly declared national parks; the drive to create wilderness areas significantly affected the abilities of local communities (First Nation peoples, newly arrived immigrants and settler communities) to meet their subsistence needs.[18] As Jacoby notes:

> For many rural communities, the most notable feature of conservation was the transformation of previously accept-able practices into illegal acts: hunting or fishing redefined as poaching, foraging as trespassing, the setting of fires as arson, and the cutting of trees as timber theft.[19]

Equally, Yosemite National Park was cleared of the Native American communities who had lived in, used and shaped the landscape for generations. Meanwhile the park was opened up to appeal to the new, white, leisured class. The classic photographs of the area taken by Ansel Adams continue to feed the myth of Yosemite as a primordial wilderness; but he deliberately framed the images so that the Native American communities never appeared, even though he had to work around them constantly.[20] The story goes that this American wilderness model was exported all over the world, first via European empires and then via the global conser-vation movement. In fact, many conservation initiatives pre-dated Yellowstone; for example, there were hunting reserves in India and China before the arrival of European colonizers, and forest reserves were established in the Roman Empire.[21] Nevertheless, whether it came from Yellowstone in the first place or not, the model of people-free wilderness contained in national parks serves the interests and objectives of the conservation movement: it allows clear and defined boundaries for their work, areas where species can be counted and biodiversity determined, areas where success and failure can be measured.[22] This has led to a simplistic view of conservation, characterised by heroes and villains. Jacoby argues:

> the history of conservation has become little more than a triumphant tale of the unfolding of an ever-more enlightened attitude toward the environment. The result has been a narrative drained of all moral complexity, its actors neatly compartmentalized into crusading heroes (conservationists) and small-minded, selfish villains (conservation's opponents).[23]

The expulsion of local people in order to create new wilderness areas in national parks is by no means past history; on the contrary, For example, the Ngorongoro Crater and Serengeti in Tanzania are important 'wilderness' areas that sustain globally important wildlife populations. Without them, we are told, we would not have the wildebeest migrations, the great herds of elephants and majestic lion prides. But this is a manufactured Eden. Maasai herders were forcibly evicted from the area to conform to a Western vision of an African paradise for wildlife. Bernard Grzimek, director of the Frankfurt Zoo and creator of the highly influential film and associated book *The Serengeti Shall Not Die*, believed that Maasai communities posed an unacceptable threat to wildlife, even though they had been using the area in a sustainable way for generations.[24] *People* simply did not fit the image of wilderness, so they were removed and excluded via a long process during which their grazing and hunting activities were progressively criminalised. Evictions and exclusions in various forms continue today. One recent example concerns the African Parks Foundation, the first private organization in Africa that takes on the actual long-term management responsibility of parks in public–private partnerships with governments, by combining 'world-class' conservation practice with business expertise. The Foundation claims that it offers an African solution to Africa's conservation challenges, emphasizing the stimulation of responsible tourism and associated private enterprise as a means by which parks can pay their way, so they can provide a foundation for sustainable economic development and poverty reduction. In its first six years, African Parks Foundation took on responsibility for managing seven protected areas in five different countries, covering a total area in excess of 2,500,000 hectares.[25]

One of those parks was Nechisar National Park in Ethiopia's Omo Valley. When the African Parks Foundation took over management of Nechisar, they effectively took over land which was traditionally the home of the Mursi community. Co-operation between the Ethiopian government and the African Parks Foundation produced a powerful alliance, and the Foundation's sudden increase in enforcement of park boundaries and regulations effectively criminalized those who lived in and around the area. Refugees International estimated that 2,000 families were removed from Nechisar when African Parks Foundation provided funding to fence the park and develop tourist facilities. The families were forced to relocate to areas outside the park boundary, creating problems in the surrounding areas which were already inhabited. The Ethiopian government claimed that the removals were undertaken 'voluntarily'; but Refugees International reported that homes were burnt down by government operatives. A request from the Mursi community prompted the creation of a new NGO, Native Solutions for Conservation Refugees, which claimed that 10,000 people were under threat of eviction. The African Parks Foundation claimed in response that the development of tourism in the area would provide hundreds of jobs for local communities, but according to their own figures only ninety local people were employed as guides, drivers and so on.[26] In a final twist, as a result of opposition to the scheme the African Parks Foundation announced that it was withdrawing from Nechisar in December 2007, citing possible legal challenges from affected peoples and the activities of human-rights organizations. The withdrawal of the Foundation also meant that their stipulation that the government remove the local Guji community was not carried out.[27]

Critics have focused much of their attention on violent evictions such as those seen at Nechisar. However, all around the world, millions of people are affected by more insidious forms of exclusion and marginalization. They are dispossessed of their lands and rights to use natural resources via conservation rules which outlaw grazing, agriculture, collection of plants for medicines and hunting for food. The drive to create wilderness areas has led to one of the murkier themes of conservation: the removal and exclusion of resident communities from protected areas.

The history of national parks demonstrates that there are two broad types of exclusion. The first is outright eviction to make way for protected areas, accompanied by forms of symbolic exclusion from the history, memory and representation of the landscape. The second kind is exclusion as a result of the loss of control over landscapes and environments.[28] This happens when communities are forbidden from entering national parks or when long-term patterns of using natural resources for subsistence are outlawed or disrupted.

An example of this type of exclusion was seen at Kumbhalgarh Wildlife Sanctuary in India, the subject of a study by Robbins et al. The resources in the sanctuary area had traditionally been used by poorer communities in times of crisis to meet their needs for food and fuel. However, in 2002 the Indian government shifted policy towards Joint Forest Management, which pushed aside decades of localized, centralized management of reserves. The shift was felt by poorer communities, whose use of resources within the parks and reserves was criminalized. However, the agencies responsible for enforcing the new rules at Kumbhalgarh did not support the new rules. Instead they resisted them through non-compliance, allowing local communities to continue with grazing, small-scale tree felling and fuel wood collection. Over time this evolved so that local communities could only negotiate access to these resources via bribes.[29]

The scale and pattern of evictions and exclusions as the first national parks were being establised are poorly documented; where they are recorded it is often in reports published in the 1990s, many decades after the protected areas were established. The greatest expansion in the numbers of protected areas was between 1985 and 1995, and most were accompanied by a series of evictions and exclusions.[30] This attempt at documenting historical grievances was in part about conservation rediscovering its dark past, and as a response to critics who complained about the catalogue of abuses of indigenous peoples and rural communities at the hands of conservation. Such evictions can result in serious economic and psychological harm.

For example, Mbaiwa's study of the development of tourism in the Okavango Delta clearly shows how local communities were displaced to make way for the Moremi Game Reserve, and the Central Kalahari Game Reserve in the 1960s. San (bushmen) communities were displaced and moved outside the boundaries of the new protected areas. They were suddenly faced with restrictions on access to important resources that they had relied on for survival. Relocation of the communities effectively ended their nomadic lifestyles, placed restrictions on hunting and introduced the idea of permanent settlement. They had to develop new livelihood strategies, including agriculture and cattle rearing, in order to survive. The promised benefits did not appear; social services and health services remain underdeveloped nearly forty years later after the initial removals. San communities have increasing felt that the Government simply handed 'their' areas over to foreign tour operators, and that the benefits of wildlife conservation were enjoyed outside the country. Even today, residents of the Khwai area in the Okavango Delta face draconian regulations implemented by the government of Botswana, which they maintain are designed to intimidate them and force them into relocating again so that the area can conform more readily to the vision of a people-free wilderness demanded by international tourists and foreign tour operators.[31]

One of the reasons conservation fails to engage local communities is because it produces and sustains criminality. Conservation agencies, NGOs and international legal frameworks define legitimate and illegitimate uses of wildlife. This is most apparent in the commitment to national parks based on a model that separates human communities from 'nature'; it is assumed that conservation is best served by the protection of wilderness areas that are people-free. But much like ideas of biodiversity, this commitment to ideal forms of wilderness is a Western fixation and is by no means a 'neutral' and apolitical position. Perhaps its two most damaging implications are the ways a commitment to wilderness implies and justifies the removal and marginalization of local communities, and how it essentially devalues the nature found outside protected areas. It assumes that nature can be saved through creating and engineering people-free environments without understanding the relationship between humans and environments or the context of economic and political forces which might drive destruction of habitats and extinction of species.[32] The authors of the Kumbhalgarh study discussed above argue that across the world resource use is governed by informal negotiations and structured by 'hidden institutions' and the ways in which rights are defined and distribute in local communities. Failure to understand this is extremely counterproductive to the interests of conservation.

This process of marginalization and exclusion is also a problem for the conservation movement's latest 'big idea': the creation of vast transfrontier conservation areas (TFCAs), sometimes referred to as 'peace parks'. These are large regional-scale protected areas that cross international boundaries to conserve entire ecosystems. One of the best known is the Great Limpopo Transfrontier Park (GLTP) in Southern Africa, which draws in territory from South Africa, Mozambique and Zimbabwe. It is based around protected areas in each of the three partner countries but has also covered land outside national parks to create 'corridors' for wildlife between the parks. In Zimbabwe the people of the Sengwe communal land were not aware of the initiative until they saw maps of the GLTP which covered their lands as well; similarly local communities in Mozambique were surprised to find themselves on the receiving end of a delivery of thirty surplus elephants from South Africa.[33] Local communities fear that they will suddenly find themselves evicted from new conservation areas, and that their attempts to use resources in the area will be redefined as criminal acts. This is understandable given the history of colonization which resulted in dispossession, and a more recent history of poor relationships with parks agencies.

Park advocates point to the benefits involved in creating wilderness and committing to conservation schemes. Tourism plays a central role in justifying conservation, and especially the need to demarcate, enforce and maintain protected areas as important 'tourist playgrounds' (the relationship between tourism and wildlife is more fully discussed in chapter six). But, in effect, local communities are excluded from protected areas and economically important landscapes used by the tourism industry. Paradoxically, tourism also opens up new areas for tourists to visit while simultaneously excluding other communities. This ensures that landscapes and wildlife habitats conform to tourist imaginings of what the 'wild' looks like, as in the case of Kruger National Park discussed earlier in this chapter. Tourists are often unaware of the social, political and economic processes that have conspired to 'produce' the attractions they want to visit.[34] They are not informed about the ways communities may have been persuaded (or even forced) to leave areas that are deemed valuable by the international tourism industry, often in conjunction with national governments and conservation NGOs. They are also unlikely to be aware of the ways in which people and communities are drawn into the tourism industry via empty promises of well-paid jobs.

GLOBAL CONSERVATION AND ENVIRONMENTAL NGOS

In 2005, Johan Eliasch, a Swedish multi-millionaire businessman, former deputy treasurer to the UK Conservative Party, and later Prime Minister Gordon Brown's special envoy on rainforests, bought up 400,000 acres of the Amazon rainforest for a reported US$8 million. His purpose was to protect the rainforest by purchasing it to conserve it; he claimed that he had made sure the area was safe from logging, and he encouraged scientists to explore it for new species and potential medicines. At first glance there seems to be nothing wrong in doing this. Eliasch's NGO, Cool Earth, has encouraged celebrities and the world's rich to join its founder in purchasing rainforests to save the earth. But by 2008, the Brazilian government was angry, and reminded potential buyers that the Amazon already had an owner, the Brazilian people. Later that year the Brazilian government fined Gethal, another company owned by Eliasch, for illegally cutting down 230,000 trees and for failing to have the correct certification for land it owned in the Amazon; the company and Eliasch claimed that the investigation and fine were politically motivated and the logging was carried out by the previous owners [35]

Cool Earth encourages supporters to 'act now' to save the rainforest and help to combat climate change. Its website contains endorsements from well-meaning public figures, including comedian Ricky Gervais and wildlife-documentary maker Sir David Attenborough. It is a perfect example of the slogan 'think global, act local': by buying something or donating money you have acted to save rainforests and combat climate change via a few clicks of a mouse and a credit card.[36] But the frustrations of the Brazilian government reveal a core problem in conservation: the ability of conservation organizations to project power across the globe, and their lack of accountability to national governments and local communities.

No discussion of conservation is complete without considering the increasing role of conservation NGOs, which now wield enormous power in the international system. NGOs filled a space in the international system which was left at the end of the Cold War. From the mid-1980s they began to get more professionalized: they hired media teams, engaged in marketing, developed large-scale workforces and focused on professional fund-raising. As the Cold War ended NGOs were catapulted onto the world stage as a key part of what was called the 'new world order'. In

1989 this new world was not defined by the old Cold War divisions between East and West; instead we placed our faith in international organizations like the UN and committed NGOs like Oxfam and WWF-International to change the world. This allowed transnational NGOs to run new global-scale campaigns about specific issues like wildlife conservation, poverty reduction and anti-globalization.[37] We trust them to really care about wildlife, to do good, to save the world. Supported by endorsements from celebrities from pop star Sting to Hollywood actor Harrison Ford, they are able to generate and channel millions of dollars in funding from well-meaning members, as well as donors, national governments, and (increasingly) private businesses. They have captured our attention, and we believe that when wildlife is in trouble, conservation NGOs can save the day.

But conservation NGOs, like other organizations, compete in a global market for our attention, and more crucially for our money. They have massive budgets, thousands of staff and projects all over the world. They have their own agendas, special interests to defend, budgets to manage and targets to meet. Their power lies in their ability to capture and disburse funding and their ability to push forward their own idea of good conservation practice. This can be highly problematic. We are told they are trying to do their best in difficult circumstances. This may well be the case, but the ways NGOs present conservation issues to the general public are often very disingenuous towards the very communities that live with wildlife. The simplistic images they create contribute to a climate where aggressive conservation strategies are not only permitted, but are applauded. Sometimes (as we will see in chapter three) they are well aware that local communities are not the real problem, but communicating the complex realities is just too difficult.

Here I want to explore the changing roles of NGOs, and look at the implications of the spectacular growth in the powers of the 'big four' conservation NGOs: The Nature Conservancy (TNC), Conservation International (CI), World Wide Fund for Nature (WWF) and Wildlife Conservation Society (WCS). Numerous global conservation NGOs are household names; NGOs like WWF, TNC and the Sierra Club have a clear international profile and they have had a critical role to play in determining how we think about conservation practice. Since the 1990s the power of the so-called 'big four' conservation organizations has grown substantially. This is due to their access to funds and to the ways they interact with powerful global networks. In 2004, the anthropologist Mac Chapin published an article in *WorldWatch* entitled 'A Challenge to Conservationists' that critiqued the concentration of funding and power in the big four NGOs. This prompted an outcry from the organizations concerned and caused a storm in the conservation community. The article and the ensuing debate highlighted the immense budgets and powers that these organizations held.

According to Chapin, investment in conservation in the South by the big four increased from US$240 million in 1998 to US$490 million in 2002, which meant that by 2002 they accounted for half of all conservation funding available for the developing world.[38] They wield immense power and claim that they alone have the scale and capacity to organize and implement large-scale conservation projects. This is very attractive to potential funders. The big four NGOs can claim that supporting and backing their schemes comes with a guarantee. For funders the attractiveness of backing big conservation organizations is that it develops an international profile which allows them to demonstrate that they are serious about doing something positive for conservation; after all, smaller donations to lesser-known organizations do not attract the same kinds of coverage in newspapers, magazines and on websites. Funders have included International

Financial Institutions (like the World Bank) and private foundations (such as the Gordon and Betty Moore Foundation).[39] It is difficult to establish exactly how much money NGOs provide for conservation projects around the world, and various studies show there is a significant degree of overlap in funding between different NGOs.[40]

But provision of funding is only one way NGOs can influence conservation practice right down to the local level. Rather like the groups of scientific experts who advise on policy directions for international legal conventions like CITES, NGOs are also important 'knowledge brokers'. Their power also comes from the ways in which they have developed a profile as a trusted authority able to tell us what needs to be saved and how it to do it. They are able to influence conservation thinking and practice through public campaigns and through the development of their own research projects on conservation which push forward and promote objectives that are in line with their mission.

NGOs have a uniquely powerful position because they are able to present themselves as committed to wildlife conservation, and therefore able to tell the truth. They can claim that they are not tainted by shackles of government or corporate thinking, and that their research, approaches and policies are neutral, scientific and in the best interests of wildlife. We are more likely to believe the information presented by WWF and CI than by national governments or private businesses, who we suspect have their own interests and agendas to protect.

NGOs challenge governments and international organizations to take action to halt species loss or manage resources more effectively. They are publicly critical of the agencies responsible for conservation. Their publicity and campaigns draw attention to specific wildlife issues and point to the ways that the public and NGOs can work together to solve complex conservation problems. NGOs have also had a big impact on global conventions, like CITES and CBD, through pressure group politics. NGOs are able to mobilize resources to produce and disseminate information through various media (television, internet, books, leaflets, adverts, magazines and reports); such campaigning activity can be used to embarrass governments and international organizations about their failings in conservation; and these activities serve to increase public awareness about particular issues like tiger poaching, or the impact of the bushmeat trade on monkeys and apes. This campaigning activity has changed the ways national government agencies, the private sector and international organizations go about the practice of conservation.

However, while we might expect NGOs to challenge, contest and criticize, they have in fact developed increasingly close and compatible relationships with international institutions, national governments and private business. This means that nowadays they are much more likely to be working alongside large corporations like Shell, BP or Rio Tinto, or international organizations like the World Bank. In 2003 CI teamed up with Starbucks, and this collaboration suits both parties: CI claims that it means local coffee growers are encouraged to conserve forests, while Starbucks is able to say it sells ethically produced coffee. CI has also worked with McDonalds to 'educate' children about the importance of conserving threatened species. This led to the development of the Endangered Species Happy Meal, and collaboration with DreamWorks Animation on *Kung Fu Panda* to educate consumers about panda conservation.[41] When DreamWorks made the animated film *Madagascar* the company made a donation to CI, which was then used to fund an ecotourism advisor to the President of Madagascar; in return DreamWorks was allowed to use the name Madagascar for its film.[42] The US-based

International Conservation Caucus Foundation links over 100 members of the US Congress with conservation NGOs like CI and major corporations like Walmart, BP and Exxon/Mobil. At their inaugural gala dinner in 2006 they honoured Harrison Ford with a 'Good Steward' Award.[43] Critics see this as greenwashing, which allows big corporations to get on with business as usual, and continue damaging the global environment under the guise of ethical consumerism.

NGOs work with such partners because they think they will be able to push their own agendas for specific types of conservation by doing a deal with donors or developers. As new dam projects, oil fields, mines and tourism facilities threaten to engulf and destroy habitats and the wildlife they contain, conservation NGOs have spotted a new opportunity to push their agendas forward. Instead of resisting these new and potentially damaging initiatives, they work with developers on 'mitigation' projects. These revolve around deals with developers to 'offset', mitigate or compensate habitat loss by paying for or setting aside land for new conservation schemes and protected areas.

The increasing concern about climate change has presented a particularly lucrative opportunity for conservation NGOs. In the emerging world of global carbon markets, conservation NGOs encourage the world's polluters to offset carbon emissions via the purchase of carbon credits, which are then used to pay poorer communities to leave forests standing, or to plant new ones to store carbon as they grow. Critics point out that this does nothing to change the behaviour of polluters, and nothing to encourage them to reduce carbon emissions.[44] It is a quick (and unsustainable) fix: pay a conservation NGO to set up a new park or plant and forest and, with conscience salved, you can carry on with business as usual. Carbon markets are new, so it is impossible to determine how much conservation NGOs will make from them;

but judging by the levels of excitement about carbon markets they are expecting to get millions of dollars. This is a very attractive option in the harsh new world of an unprecedented global financial crisis which threatened to damage the coffers of our best-known wildlife NGOs.

Michael Goldman's study of the development of the Nam Theun 2 Dam in Laos as part of the Mekong River Basin Development Programme in South East Asia demonstrates these new patterns of co-operation between conservation NGOs and the very organizations we would once have expected them to campaign against. The development of dams in Laos drew together an unlikely configuration of organizations: the Swedish International Development Cooperation Agency (SIDA), IUCN, WCS, the Asian Development Bank and the World Bank. Because of the history of international criticism of World Bank involvement in large dam projects, the Nam Theun 2 plans involved environmental and social impact assessments and planning right from the start. Interestingly this meant that the World Bank and the Asian Development Bank joined with some of the world's biggest conservation NGOs. The dam project contains a series of environmental projects including biological corridors for wildlife, watershed conservation and ecotourism projects, but also health clinics, workshops on birth control and experimental farms. Conservation NGOs have joined with international institutions to develop the dam project, and act as 'mitigators' of the social and environmental impacts.[45] Local communities have been redefined as specific kinds of user groups rather than as inhabitants: as non-timber forest product users, poachers or slash and burn agriculturalists. The ways these communities are defined then allows a whole series of interventions and policy frameworks drawn up by the agencies involved in the Mekong Basin project.

These interventions aim to turn local communities in to 'good conservationists' who can manage the environmental changes in line with the vision of external organizations. The purpose is to recreate people in ways that make them fit better with the interests of conservation NGOs (and increasingly their business partners). In the case of the Nam Theun 2 Dam this included the development of new rice-growing areas because of enhanced irrigation, which encouraged local communities to change into commercial rice growers. Using a series of incentives and developments associated with the dam, including new conservation rules, the indigenous communities were transformed into a new workforce for agriculture and hydroelectric plants. This was justified as a good development for conservation, with claims that it was necessary to prevent poaching and the 'environmentally destructive' practice of slash and burn.[46]

This kind of social engineering in favour of conservation is not restricted to the Mekong Delta. Tania Murray Li agues that conservation NGOs have been engaged in programmes motivated by a 'will to improve'. In this sense they are the latest round of external intervention, which started with missionaries and colonizers that viewed local communities as objects that must be improved for their own good – and now for the global conservation good. In Dongi Dongi in Indonesia, local farmers believed that outside agencies, including the government and TNC, were keen on developing a national park so they could demarcate it for mining or for extraction of genetic resources. They resisted the development of the park by occupying the area. Meanwhile the government and TNC justified plans to evict people from the park on environmental and humanitarian grounds. Authorities claimed that continued human settlement in the park damaged watersheds, increased the risks of siltation and flooding, and had the potential to threaten the livelihoods of 300,000 people who

ownstream communities. Local farmers disputed this and
the government for choosing steppe areas for resettle-
.[47]

...... played a critical role in shaping and recreating local
people so that they behaved in ways that were compatible with
TNC's vision of good conservation. In a departure from the
Integrated Conservation and Development Project (ICDP)
model of the 1990s, TNC did not attempt to alleviate poverty in
park border villages, or provide economic compensation for loss
of access to park resources. Instead they tried to change people's
expectations, desires and practices so that they would 'choose' to
engage in park protection. TNC called the approach 'collabora-
tive management', which Li suggests was an uneasy blend of
co-operation and betrayal. Yet TNC declared its approach as
a major breakthrough because it persuaded local farmers to
develop new aspirations that were in line with TNC's vision of
biodiversity protection.[48]

Conservationists cast themselves as engaged in progressive
projects, underpinned by good intentions. But this exposes them
to accusations of interference and neo-colonialism. Their ideas
and policies are often inspired by Western models of nature
conservation, and when they are introduced at the local level they
are alien and unwelcome. This results in the imposition of
specific ideas of nature and conservation, but also can mean the
physical displacement of poor people to create protected areas
for occupation by the international tourism industry, research
scientists and conservation NGOs.[49]

The growing shift towards co-operation between NGOs, the
private sector, the World Bank and governments means that
NGOs have a great deal of input into defining what constitutes
good conservation practice at the national and local levels. This
power is magnified through their increasing willingness to work

with other international institutions like the World Bank. In effect, NGOs can increase their chance of getting national governments to sign up to their vision for conservation because they are able to control large amounts of funding and are able to offer a wider range of resources, which are increased by their involvement with the World Bank and others. In return, institutions with a 'bad press' like the World Bank get their green credentials enhanced because they are able to advertise the fact that they are working in conjunction with household 'green names' like WWF and others. The problem is that conservation NGOs lack a critical independent voice, and instead have to satisfy their partners, be they donors, international institutions or corporations.[50]

NGOs are not just co-operating with powerful global institutions, they are transforming the ways national governments make policy in the realm of conservation and in broader questions about economic development. Even so, these new patterns of co-operation result in familiar patterns of displacement, marginalization and exclusion of poorer communities via the development of new conservation rules. One recent example comes from Madagascar, where the close relationship between conservation NGOs, the World Bank and global development donors was used to create a new network of protected areas.[51] The high profile of international environmental NGOs in Madagascar is unusual. Madagascar is well known for high levels of biodiversity and endemic species but suffers from severe environmental problems.[52] As a result numerous global donors and conservation NGOs have identified it as a high priority.[53] The main players involved in Madagascar were global environmental NGOs, especially WWF, WCS and CI. They linked up with the World Bank and donor agencies of various wealthy countries, especially the United States Agency for International Development (USAID), Cooperation Suisse and Cooperation Française.

This co-operation was formalized as a donor consortium which worked with the government of Madagascar as one of the partners. Its key member was the World Bank, but it also included USAID, the German government, the Japanese government, the French government (Cooperation Française); the Swiss government (Cooperation Suisse), CI, the WWF and the WCS (the latter joined in 2004). The consortium met monthly to review the progress, and determine future funding priorities and policies for Madagascar. Its composition highlighted the unique nature of politics in Madagascar; nowhere else had three international wildlife NGOs directing national policy.

The power of NGOs was evident: they successfully persuaded the Malagasy government to increase the number of protected areas. In 2003 President Ravalomanana committed Madagascar to tripling the amount of land with protected area status within six years to create a six million hectare network of terrestrial and marine reserves.[54] The commitment was named the 'Durban Vision Initiative', because it was announced by the President of Madagascar at the 2003 World Parks Congress held in Durban, South Africa. Within Madagascar, it led to the creation of the 'Durban Vision Group', which included donors, NGOs and Malagasy government agencies. WCS and CI argued strongly that the initiative was agreed in consultation with the Malagasy government and that Malagasy organizations would be partners or stakeholders in it.

By 2007 CI was pleased to report the creation of new protected areas that covered three large areas of land: the 499,598 hectare (1,929 square mile) Fandriana-Vondrozo Forest Corridor in the south-east; the 276,836 hectare (1,069 square mile) Mahavavy-Kinkony Wetlands Complex of lake, river and forest on the north-west coast; and the Menabe Central Forest, 125,000 hectares (483 square miles) of dry deciduous forest in the south-west. A number of other smaller protected areas had been declared to create corridors for wildlife between existing national parks. CI had also set up the Global Conservation Fund (GCF) and had provided US$2.2 million in support.[55]

As soon as the Durban Vision Initiative was announced there were complaints that WCS and CI had placed an enormous amount of pressure on Ravalomanana to develop a conservation policy in line with their mission for biodiversity conservation.[56] The NGOs were able to play on the fact that Ravalomanana was a new president, and had taken over from the corrupt regime of Didier Ratsiraka in 2003; at that point Madagascar was an extremely poor country with few offers of donor support. Ravalomanana was clearly hoping that the USA would become the main donor to Madagascar, to reduce dependence on the former colonial power, France. Critics have suggested that he had felt obliged to agree because of threats from the NGOs that they could effectively lobby in Washington to reduce US support to him. Other members of the donor consortium expressed worries about the powers of CI and WCS. There were serious concerns that the commitment to new protected areas (no matter how community-oriented and participatory they might seem from the outside) sent a message that wildlife and habitats were more important than people; and that they could result in forced evictions from the newly declared protected areas. It seemed that the government had contracted out conservation in Madagascar to external conservation NGOs.

The case of Madagascar demonstrates that conservation NGOs are powerful global actors which can direct and determine conservation policy at a national and local level. They are heavily involved in determining what we understand as legal and appropriate use of wildlife versus illegal and inappropriate use. The ways they define good conservation practice produces and sustains criminality. Conservation rules create categories of criminals, villains, law-abiding citizens and wildlife saviours.

CONCLUSION

The global framework for conservation is vitally important, and helps us understand how some activities come to be defined as criminal and illegitimate while others are considered legal and perfectly appropriate. Global conservation has two forms: the first is global legal frameworks and the second is the ways NGOs inform and shape conservation practice. Global legal frameworks like CITES and the CBD provide us with the definitions of what is legal, what is appropriate and what is sustainable use of wildlife; they also tell us what is illegal, what is criminal and what is illegitimate. But these definitions are not neutral scientific decisions: they are produced by political dynamics and by moral judgements about the uses of wildlife. NGOs have an important role to play in informing and shaping conventions. Conservation NGOs are trusted as organizations that are committed to saving wildlife, that are 'above politics' and working in the true interest of the environment. They are uniquely placed to produce knowledge about wildlife and the impact of criminality, and they are highly skilled at campaigning and communicating their standpoints. But they are highly political organizations, and are able to shape and define how we go about conserving wildlife.

It is important to remember that NGOs do have their own agendas, and are in competition with each other for funding, attention and public support. Both Conservation International and Wildlife Conservation Society are prominent advocates of a return to fortress-style approaches to conservation which has begun to displace the idea of more community-oriented approaches. Fortress-style conservation is best described as the 'fences and fines' approach where national parks are fenced in, local communities kept out and any trespassers met with fines, or worse with shoot-to-kill policies.[57] This produces a highly counterproductive strategy, responsible for serious human-rights abuses and social injustices against the poorest communities in the world. This is the subject of the next chapter.

CHAPTER THREE

WILDLIFE WARS:
POACHING AND ANTI POACHING

IN MAY 2008 I was in the Okavango Delta in Botswana. We picked up a woman and her grandchild to give them a ride from the town of Maun to their settlement near village of Ditshiping and the Daunara veterinary fence checkpoint. They lived in an area known as NG32, half of which was leased to two safari lodges for photographic tourism and to a sport hunting operator. In the middle of nowhere, the woman insisted that we stop: she had spotted a dead honey badger lying by the roadside. Closer inspection revealed it had been freshly killed by a passing vehicle. The woman wanted to take it home to cook and eat. However the driver and guide (who worked for a safari operator) were insistent that while they were sympathetic to her request they could not risk taking the dead honey badger in the vehicle; they argued that if they were intercepted by the parks service they would be accused of poaching and all of us would be arrested. The woman protested that the honey badger was already dead, so to leave it would be a waste of good and much-needed meat.

But we had no way of proving that we had not deliberately killed the honey badger, and so it was left by the roadside to rot in the sun.

The honey badger is unlikely to excite international tourists wanting to see and photograph the 'big five' (that is, the lion, African elephant, Cape buffalo, leopard and rhinoceros), nor is it likely to be a popular sport hunting trophy; but the local communities in the area had signed up to a 'no hunting' rule, and in return they were promised benefits from tourism. It is important to remember here that we were not inside a national park, or even a buffer zone around a park where activities which might affect wildlife and habitat conservation are often outlawed. Instead, we were in an area owned by the local community but leased to private safari operators. The operators had insisted that hunting for small game had to stop in order to protect the wildlife stock for wealthy foreign tourists to shoot with a camera, or with a gun. These rules were being enforced by the government's own parks service on behalf of the private safari operator. This story highlights the problems posed by the rights of rural communities to use wildlife in their area (dead or alive) for their own subsistence purposes; it reminds us of the difficulty of distinguishing between poaching and hunting. It brings the issue of social justice into sharp focus: in this case it appeared that because the communities had signed away their rights to hunt, the interests of tourists, sport hunters and safari operators took priority.

International conservation organizations such as CI, WWF and WCS have played an important role in defining understanding of good conservation practice and of why and how poaching occurs. But the complex problems in defining poaching versus hunting are rarely reflected in the campaigns by conservation NGOs and claims by national governments that they are engaged in a 'war' to save species from criminal gangs of poachers. For example, IFAW states 'exploiters often have the upper hand. Under-equipped and under-staffed African wildlife law-enforcement agencies are easily outmanoeuvred in their attempts to fight poaching by sophisticated crime syndicates operating freely across national borders'.[1]

EIA provides the following definition of tiger poachers: 'sometimes they are local criminals, known poachers prepared to flout the law for the promise of high rewards. At other times they are disgruntled locals, fed up with tigers killing their livestock or wild deer eating their crops.'[2] Images of elephants and rhinos with their tusks and horns hacked off and covered in blood are used to shock the public into supporting military style anti-poaching initiatives to protect wildlife. But anti-poaching raises a series of questions: if we accept that it is intended to protect wildlife, then it is important to consider who we are protecting wildlife from, what we are protecting wildlife for and what methods we deem acceptable. The answers to these questions lie in the ways we think about why poaching happens and who actually carries out the poaching. It is vitally important to understand the dynamics of poaching in order to offer a long-term solution to the problem.

This chapter will look at some important questions in conservation: how and where we draw the line between hunting and poaching, the differences between subsistence and commercial use of wildlife, and how legitimate and illegitimate uses of wildlife are defined. It will examine anti-poaching strategies and the problems they pose over who gives permission for the use of violence to protect wildlife, and whether conservation ends justify violent means, especially in the use of shoot-to-kill policies in protected areas.

POACHING VERSUS HUNTING

International campaigns give the impression that parks agencies and conservation NGOs are engaged in a continual battle to protect wildlife from armies of highly organized poachers who are motivated by greed. The war to save the planet is commonly presented as a just war, motivated by the highest morals and waged by dedicated conservationists. But if we examine the idea of a war to protect wildlife more carefully, it is clear that it is used to justify highly repressive and coercive policies against the world's most marginalized and vulnerable people. The war for conservation carefully privileges the need to save wildlife over the rights of people who live with it.

Since the 1980s conservation practice has been progressively militarized following high-profile ivory and rhino wars in East Africa. Governments responded to international pressure to save elephants and rhinos through the development and deployment of military-style practices. For example, in 1988 President Moi in Kenya gave permission for the Kenya Wildlife Service to use a shoot-on-sight policy against suspected poachers; in 1987 President Mugabe in Zimbabwe declared a war on rhino poachers and gave authorization for the Parks Department to engage in a shoot-to-kill policy; and in Tanzania Operation Uhai in 1989 was used to sweep all parks and neighbouring communities of suspected poachers.

The shoot-to-kill approach was not confined to the 1980s: parks agencies still use deadly force for conservation purposes, with or without state authorization. For example, in just two years (1998–2000) parks staff in Malawi, who had been trained by South African mercenaries, were implicated in 300 murders, 325 disappearances, 250 rapes and countless instances of torture and intimidation in the Liwonde National Park alone.[3] These strategies have been supported and funded by international NGOs and their memberships through sponsorship of rangers, equipment and training. It is clear that the use of deadly force has become commonplace, especially in Sub-Saharan Africa where international pressure to save high-profile species is intense and where the development of nature-based tourism has meant many of these species are definitely worth more alive than dead.

However, a closer examination of poaching practices and the dynamics that drive and sustain poaching reveals a very complicated picture. Poaching actually covers a wide range of activities, and how we define it is a key question. At its most basic poaching is illegal hunting of any animal, whether banned by the state or by private owners of wildlife. While poaching of well-known and well-loved species like rhinos, elephants and tigers makes the international headlines, poaching covers a much wider range of animals and activities. In general, there are two types of poaching – subsistence and commercial – but there is a third kind which is discussed later in this chapter: poaching as a conflict strategy, where wildlife might be used to subsidize armies, rebel groups and militias or where killing of animals is used to send a message of resistance. Wildlife agencies are faced with the difficult task of tackling poaching, and in order to understand the ways parks departments and NGOs approach poaching it is worth exploring the different kinds of poaching a little further, since the ways in which poaching is defined and understood have a direct relationship to the ways conservation agencies design and implement anti-poaching strategies.

How Hunting Became Poaching

It is tempting to think of the apparent mass destruction of wildlife through poaching as a problem that appeared in the late twentieth century. But in fact the history of the phenomenon is important for understanding how current definitions of poaching versus hunting were developed and then consolidated into current practices.

Poaching for bushmeat and tiger parts have been the hot topics of the early twenty-first century, but wildlife has a long history of being used to subsidize conflicts and imperial expansion. Ivory poaching excites the public imagination more than any other issue, but there is little appreciation of the wide and varied dynamics which drive and sustain hunting elephants for their ivory. Ivory hunting has a very long history, especially in Sub-Saharan Africa; ivory was a key product in trading links between the interior of the continent and the trade routes of the Atlantic and Indian Oceans.

Colonial expansion was initially underpinned by the search for lucrative goods including gold, ivory and slaves. John MacKenzie's book *Empire of Nature* details how hunting for game assisted and subsidized British imperial expansion in Africa in the late nineteenth and early twentieth centuries. The (in)famous hunters of the Victorian era, such as Frederick Courtney Selous and Lord Lugard, used trading in ivory and hides to subsidize their journeys, and killed wildlife to feed expedition staff.[4] Hunting of wildlife by European explorers and settlers in British colonies in Africa was extensive enough to cause concern in London, where the Society for the Preservation of the Wild Fauna of the Empire was established in 1903.[5] In the early part of the twentieth century hunting expeditions by high-profile 'celebrities' of their day, such as Theodore Roosevelt, served to bring the idea of the safari to international attention. Roosevelt's tour of Africa in 1908 was indicative of a new wave of safari hunters, with the result that by the end of the First World War safaris had become a specialized international business. Soon big game hunting safaris were a mainstay of the annual itinerary of wealthy elites, to be fixed in the popular imagination as a vital leisure pursuit by Hemingway's *The Snows of Kilimanjaro* (1936).[6]

Wildlife also acted as an economic underwriter for imperial expansion. In East Africa the lure of ivory drove European expansion, and as people became wealthier in Europe and North America the demand for ivory products such as jewellery, piano keys and billiard balls grew. This is not a story that is unique to the British Empire and to Sub-Saharan Africa. We are all familiar with the story of the dodo, and perhaps less so the giant aldabran tortoise; both species were driven to extinction on Mauritius by Dutch sailors using them as a free source of meat during stop-offs on long sea voyages. In Galapagos, the giant tortoises were also driven to the brink of extinction as a result of exploration, imperial expansion and global trade. Once it became clear that the tortoises could live for up to a year without food and water, they were taken on board ships to be slaughtered on long journeys as a source of fresh meat. Without these 'wildlife subsidies' imperial expeditions to lay claim to new lands and frontiers would have been far more costly.

In contrast to their own activities, colonial authorities often outlawed hunting with the use of snares and traps, which effectively criminalized the subsistence hunting practices of local communities. For example, in British colonies hunting using traps and snares was outlawed. In effect this meant that hunting by African communities was instantly redefined as a criminal act (poaching) while hunting for sport, leisure and trade by Europeans was defined as legal and acceptable. European hunters were presented as sportsmen engaged in noble activities; it could be said that they were proto-conservationists because they wanted wildlife to be conserved so they could hunt it for pleasure. In addition, European sport hunters were also portrayed as caring about the suffering of individual animals, amid claims that it was more sporting to ensure a quick kill so that the animal did not suffer; in contrast African hunting methods that relied on traps and snares were redefined and presented as cruel and unsporting, leading to a long and painful death.

The characterizations of hunters versus poachers also linked in to racial stereotypes of the day. Representations of African men as cruel and greedy poachers neatly intersected with European/imperial fears of Africans as savage, uncivilized and barbaric. These stereotypes are still deployed today in debates about wildlife conservation. White men have been portrayed as sport hunters and committed conservationists who were keen on defending wildlife, black men as greedy and cruel poachers, blamed for destroying wildlife.[7] This brings us back to the question of why some uses of wildlife are permitted while others are not. How we draw the line between hunting and poaching is still informed by the ideas and rules draw up by colonial authorities. The ways in which colonial authorities defined European hunting with guns as legitimate, but hunting with snares by African communities as criminal, continue to define our understandings of poaching. Yet the historical background to thinking about how and why some communities 'poach' while others 'hunt' is often forgotten in debates about how to achieve conservation.

Subsistence Versus Commercial Poaching

The practices defined as poaching by conservation agencies actually cover a variety of activities. They include fishing, hunting small game and collecting birds' eggs, as well as hunting for lucrative products like ivory and tiger bone. Here I explore the range of activities which fall under the definition of poaching. What is clear is that poaching is part of a vast industry interlinked to global markets on wildlife products and encompasses a number of different types of poachers, ranging from local subsistence poachers to ruthless commercial poachers allied to governments and illegal smuggling rings.

Subsistence poaching is commonly thought of as 'hunting for the pot', where rural communities hunt small amounts of wildlife for food; good examples of this are when fishermen in Guatemala hunt manatee (sea cows/dugongs) or villagers in Zambia hunt impala for food. Subsistence poaching is used only to meet the food needs of local communities and is usually a continuation of a historical practice of using certain types of wildlife for food. It relies on the use of low-technology traps and snares, so it tends to be quite small-scale and has a minimal impact on wildlife populations when compared with larger-scale commercial poaching. However, such hunting became redefined as the criminal act of poaching with the dawn of the colonial era, so that even today often the only legal rights to hunt are held by wealthy European and American sport hunters on trips organized by international safari businesses. Local use of wildlife, especially in protected areas, was outlawed by colonial administrations and this continued in the post-colonial period, in part because the international conservation movement pressured the newly independent governments to maintain and even expand national parks systems.[8] As a result subsistence hunting remained a criminal act and so subsistence poachers continued to be treated in exactly the same way and face the same penalties as commercial poachers, even though their motives and impacts were vastly different.

In contrast, commercial poaching is often presented as a tale of ignorance, greed, mass killing of wildlife and linkages to the international illegal trade. As with other forms of illegal hunting of wildlife, commercial poaching is a complex phenomenon. Commercial poaching of elephants for ivory has perhaps received the greatest levels of attention, and much of chapter four is devoted to the controversies over the ivory trade. However, a wide range of species are poached for commercial purposes because they are hunted for valuable body parts which are then traded internationally. The methods used for hunting also differ – rather than snares and traps, commercial poachers are much more likely to be heavily armed with automatic weapons to ensure a greater number of quicker kills. In 2007 EIA claimed that organized gangs were moving into environmental crime, because it was low risk but high value. They have lobbied the UN Convention Against Transnational Organized Crime and the United Nations Office on Drugs and Crime (UNODC) to take wildlife crime seriously and treat it as on a par with drug and arms trafficking.[9]

Tiger poaching is a good example of this kind of commercial-scale hunting. As discussed earlier, there are two basic reasons why the number of tigers in the wild has dropped. The first is habitat loss, which is usually explained as a result of expanding human populations and conversion of land for agriculture to feed growing communities. But it is quite clear that in many tiger areas their habitats are not being converted to grow food for expanding human populations; instead their habitat is disappearing under palm oil plantations, which serve global markets as an ingredient in processed foods and the growing biofuels industry. This habitat loss is compounded by an increase in conflict between humans and tigers as the animals stray outside parks and into villages and fields in search of food. This means that tigers are killed in retaliation for attacks on people and livestock. The second reason for their decline is poaching to feed the demand for tiger skins and body parts used in Traditional Chinese Medicine (TCM), which is covered elsewhere in this book; for now it is clear that despite the international ban on trading tigers and their body parts under CITES, poaching and trading of tigers continues.

WWF reported that a raid in December 1999 in the Indian state of Uttar Pradesh resulted in one of the largest seizures of body parts of big cats, including the skins of four tigers and seventy leopards, and 18,000 leopard claws. In 2008 India's National Tiger Conservation Authority (NTCA) announced an official estimate of 1,411 tigers surviving in the wild in India; this figure was more than 50 per cent lower than the previous census in 2001–2, which estimated that 3,642 were left in the wild. A TRAFFIC report in 2008 claimed that tiger body parts were on sale in 10 per cent of the 326 retail outlets surveyed during 2006 in twenty-eight cities and towns across Sumatra.[10] Despite an international ban on trading tiger parts, there is still a demand for their teeth, claws, bones and skins.

The continuing illegal trade in tiger bones and skins reveals the ways that poachers are interlinked through an organized illegal system of middlemen and traders who have the capacity to move the tiger parts from the places they were hunted to the places where they are finally sold. For example, in 2007 Indian authorities announced that they had successfully broken a smuggling ring which bought and sold skins, including tiger pelts. The traders would pay US$4,500 per tiger pelt, an enormous sum to the Indian villagers who actually hunted the animals; but the skins would eventually be sold in China for US$50,000, making a huge profit for the traders.[11] The lack of enforcement means it is highly unlikely that traders will be caught. For both hunters and traders it is a high-value/low-risk activity, and this makes it very attractive.

The Wildlife Protection Society of India has also identified the market for tiger skins in Tibet as a driving force for poaching in India. The traditional robes worn by Tibetans, *chubas*, were increasingly being made from or trimmed with fur from big cats. The trade into Tibet from India was highly organized by networks of middlemen and merchants who were able to smuggle the skins across international borders from India, through Nepal to Tibet. The growth of a more affluent and urbanized middle class in Tibet fuelled a rise in demand for the robes, and the Tibetan capital Lhasa gained a reputation for being *the* place for buying exotic cat skins. This reputation drew in illegal traders from China and Europe looking to buy tiger and leopard skins. However, it was not enforcement measures that dealt the significant blow to the Tibetan fur trade, but the Dalai Lama. In 2006 he urged all Tibetans to stop wearing furs and said he was ashamed of Tibetans who wore tiger skins, because this was against Buddhist principles. The Dalai Lama's remarks resulted in a wave of bonfires across Tibet. Although the bonfires were built to destroy the tiger furs, they were regarded with suspicion by Chinese authorities, who were concerned that the bonfires were a political statement against Chinese rule rather than anything to do with conservation.[12]

Clearly tiger poaching is a complex phenomenon: it has many different causes and is driven by demand in a variety of markets, ranging from TCM to traditional robes. Despite these variations, tiger poaching tends to be presented in very simplistic and one-sided ways. When government parks agencies and international NGOs present us with a picture of a greedy and cruel poacher, the image of the individual poacher is completely stripped of all the complicated dynamics that make tiger poaching an attractive option. We are encouraged to believe that poachers are common criminals, usually serving their own selfish interest rather than meeting the demands of a global market place driven by millions of consumers. But we cannot understand individual poachers without considering the global context in which they operate.

Mixing Subsistence and Commercial Poaching

Recently some subsistence poaching with snares has mutated and changed so that it is now more difficult to distinguish it from much more destructive commercial-scale poaching. As the demand for so-called bushmeat has grown, subsistence poaching to meet local needs has been transformed into a bigger industry. The focus here is not on animals that have traditionally been poached for lucrative body parts, such as tigers or elephants; usually the animals hunted for the bushmeat trade are apes, monkeys and antelopes which have traditionally been used as sources of food by rural communities. But this traditional small-scale use of wildlife has been transformed again by the demands of a global marketplace for meat, tropical timber and key minerals.

The growth of the bushmeat trade in Central and West Africa is linked to the extension of logging and mining into remote areas of rainforest. Whereas rainforest communities may have hunted wildlife for food on a small scale, the arrival of logging and mining companies with a large workforce to feed means the market for bushmeat has exploded. It is not just the loggers and miners that are driving demand; mines and logging camps require development of roads to transport timber and minerals to urban centres and international markets. Therefore, new markets for bushmeat have opened up in urban areas and in international trade because the meat can be transported with relative ease. In some areas this means that hunting by local communities has been displaced by organized commercial interests hoping to bag wildlife to meet demand from cities and international diaspora communities. Furthermore, hunting for export to meet the demands of the African diaspora has also increased; for example London has now become a major market for bushmeat from West Africa. The UK government has expressed worries about the amount of bushmeat being smuggled into the country: in 2007 the meat of zebra, anteaters, and several species of monkey were smuggled into the UK.[13]

In the Democratic Republic of Congo (DRC), the pygmy chimpanzee – the bonobo – has been the focus of fears for WWF and demonstrates how commercial bushmeat trading and subsistence hunting are increasingly entangled with one another. In 2004 a WWF team found that around 10,000 individual bonobos remained, much lower than the expected 50,000. The blame was placed on the long-running civil war which had driven people to use bushmeat to feed themselves, and militias to use bushmeat to feed their armies.[14] A TRAFFIC Report in 2008 revealed that conflict in the central African region was driving people in refugee camps in East Africa to hunt bushmeat to feed themselves; the bushmeat was delivered and cooked at night to avoid detection, and was referred to as 'night-time spinach'. TRAFFIC accused the agencies that ran the refugee camps of turning a blind eye to the poaching, which they claimed was caused by a lack of protein in the rations the agencies gave out.[15] Although refugees were using the wildlife to meet their food needs, the large settlements in areas where people could not grow or hunt their own food made them dependent on markets in bush-meat. It is very clear that over the last ten to fifteen years some forms of subsistence hunting have been transformed into commercial-scale hunting for profit rather than to meet everyday food needs.

Growing international concern about the bushmeat trade led to the creation of the Bushmeat Crisis Task Force, a group of organizations all concerned about the impact of bushmeat hunting and trading. It includes a diverse range of organizations including Conservation International, Dian Fossey Gorilla Fund International, WWF-USA, Toronto Zoo and the American

Association of Zoos and Aquariums. The Bushmeat Crisis Task Force has stated:

> Commercial, illegal and unsustainable hunting for the meat of wild animals is causing widespread local extinctions in Asia and West Africa. It is a crisis because of rapid expansion to countries and species which were previously not at risk, largely due to an increase in commercial logging, with an infrastructure of roads and trucks that links forests and hunters to cities and consumers. The bushmeat crisis is a human tragedy as well: the loss of wildlife threatens the livelihoods and food security of indigenous and rural populations [which] most depend on wildlife as a staple or supplement to their diet, and bushmeat consumption is increasingly linked to deadly diseases like HIV/AIDS, Ebola, and Foot and Mouth disease.[16]

The rise of the bushmeat trade has changed the nature of subsistence hunting in many areas of the world, with the result that it has displaced and overtaken traditional hunting for the pot. Where once local communities hunted small amounts of wildlife for food, the opening up of roads to new markets in logging camps and urban areas means it is more profitable to hunt for sale rather than to eat. Over time organized groups of poachers have moved in, hunting out forested areas so that wildlife is no longer available for rural communities to use as food. Therefore, the Bushmeat Crisis Taskforce has identified the trade in bushmeat as damaging abilities of rural communities to feed themselves, increasing poverty and malnutrition. Furthermore, there are real concerns that consumption of meat from monkeys and apes may be a future source of epidemics, since meat can be the means by which deadly diseases 'jump species' from wildlife to humans.

The Political Dynamics of Poaching

Poaching is not just about global chains of production and consumption: it is also entangled with the wider political dynamics of conflict. The broad political context of poaching clearly demonstrates how poaching and individual poachers operate to serve the demands of bigger networks and global markets. The 1970s and 1980s saw the development of large-scale commercial ivory poaching to satisfy demand in Europe, North America and the increasingly wealthy Asian region, especially ivory markets in China and Japan. Even though ivory could be legally traded, the legal supply could not meet rising demand, fuelling a lucrative illegal trade in poached ivory. The demand for ivory became so great that poaching halved Africa's elephant population in just twenty years.

In 1988 the African Wildlife Foundation (AWF) blamed the 'slaughter' on the insatiable greed of ivory hunters.[17] But this characterization was unsophisticated and failed to recognize the complexities involved. It communicates a clear idea that elephants could be saved by stopping greedy poachers with effective anti-poaching patrols. This idea proved to be bewitchingly simple and was a message that the general public, international donors and conservation NGOs could get behind and support. However, more recent work on ivory poaching demonstrates that it is a complex phenomenon and that large-scale poaching as witnessed in East Africa in the 1980s or parts of Southern Africa in the 1970s and 1980s could not possibly have been carried out without corruption and complicity at the highest levels of government.

In Southern Africa, large-scale poaching was part and parcel of the military campaigns in the region. One of the best-documented cases is the apartheid regime in South Africa's use of hunting and trading of ivory, rhino horn, hardwoods and drugs to fund its wars and destabilization campaigns in South West Africa (now Namibia), Angola and Mozambique. In 1995 Mr Justice Kumleben was appointed to head a commission of enquiry into the role of the South African Defence Force (SADF); the report, published in 1996, exposed massive level of SADF involvement in poaching in the Southern African region during 1980s. It was made clear that the apartheid government used ivory and rhino-horn poaching and smuggling as part of its programme of desta-bilization of states that had gained independence and moved to black majority rule (the so-called Front Lines States) in the 1970s and 1980s.[18] Destabilization involved the use of military and economic weapons against the Front Line States which aimed to create a fundamental shift in their policies towards apartheid, and this was to be achieved by attacking neighbouring states in areas where they were vulnerable. One area of vulnerability was protection of endangered species from poaching.

The Kumleben Commission found that the SADF used these sales to finance numerous rebel groups including UNITA in Angola and RENAMO in Mozambique (this had been vigorously denied in the 1980s and early 1990s, when investigations were reported to have been blocked by senior officials),[19] and that the Military Intelligence Department was heavily involved in the sale of elephant tusks from Angola between 1978 and 1986.[20] According to a high-ranking South African military officer, Colonel Jan Breytenbach, certain elements in the Chief of Staff (Intelligence) had used a front company, Frama Inter-Trading, to clandestinely export teak, ivory, gems and drugs from Angola, and with the aim of financing UNITA and South African operations.[21]

Poaching, or illegal hunting of wildlife, can thus be driven and sustained by the wider dynamics of armed conflict and disputes over the use of natural resources. Poaching is not a simple phenomenon, but involves many different kinds of people, practices, organizations and networks. It is caused by a wide range of factors: the need to eat, the desire to make money, even as a military strategy to destabilize an entire region. The international media and global NGOs tend to focus on poachers as individuals rather than delving into the basic reasons why poaching occurs and how the poached goods are traded in the international system. This 'depoliticizes' the issue: it is easier to demonize villagers as greedy or patronize them as impoverished, but far more difficult to talk about why poaching wildlife is an attractive option, to investigate the major figures in wildlife smuggling, or to draw attention to the role of international demand and markets in wildlife products. This produces a real problem: it means that anti-poaching activities are aimed at the symptoms (the poachers) rather than the causes of poaching (global demand, the international dimensions of warfare, displacement of people to refugee camps). Anti-poaching strategies overlook the role of international smugglers, whose activities are somehow divorced from the actions of the poachers.

FIGHTING THE 'WILDLIFE WAR'

Anti-poaching strategies closely follow the ways that poaching is defined. Since the mid-1980s anti-poaching has become increasingly militarized because of the idea of a legitimate and 'just' war on poachers. The use of military language and tactics in conservation is not necessarily new. Early game wardens in British colonial administrations were often ex-military personnel; in 1952 Kenya's parks staff assisted the colonial government in suppressing the Mau Mau uprising.[22] However, in more recent times the idea of a 'war to save biodiversity' has taken hold, and as a result conservation organizations and their supporters have increasingly backed military-style campaigns to prevent poaching, even encouraging the use of shoot-to-kill policies. This is especially the case in

Sub-Saharan Africa, where governments and conservation NGOs have issued orders that Africans found in national parks without permission can be shot on sight. In the fear-laden vision of a war for biodiversity, colonial definitions of hunting versus poaching have persisted and justified highly coercive and violent responses to poaching. This military-style response has increasingly displaced more community-oriented and participatory approaches.

In the 1990s, Community-Based Natural Resource Management (CBNRM) was backed by global conservation NGOs such as WWF, by the World Bank and by government wildlife agencies as *the* answer for conflict between conservation and poverty alleviation. CBNRM, an incentive-based scheme which relies on creating local support for conservation and for anti-poaching patrols, seemed to offer a win–win scenario whereby local communities could benefit from wildlife conservation. Anti-poaching campaigns follow on from the ways poaching is defined. Hence CBNRM and similar schemes are often used where subsistence poaching occurs, or where wildlife authorities assume poaching is undertaken to meet basic economic needs. They are intended to compensate local communities for giving up their rights to use wildlife in protected areas, and usually take the form of distribution of wildlife meat from culls of surplus animals or promises of employment in the national parks as rangers or tour guides. However, this approach was persistently criticized as incapable of tackling poaching in the long term, especially commercial-scale poaching for lucrative wildlife products like ivory, tiger bone and rhino horn. CBNRM certainly gave many communities an incentive to conserve wildlife. But it was increasingly criticized for being top-down and encouraging communities to act as the eyes and ears of the state in rural areas.[23]

A recent article by Hutton, Adams and Murombedzi noted a turn away from CBNRM and a resurgence in what is characterized as

3.1 Community-based tourism scheme: polers and canoes in the Okavango Delta, Botswana (2008)

'fortress-style' approaches to conservation: the 'fines and fences' approach where national parks are fenced in and patrolled by armed rangers, while poachers are fined and face jail or even death.[24] In 1999 John Terborgh published *Requiem for Nature*, in which he

controversially called for the creation of an international conservation police force to enforce internationally important protected
areas. In the book, he criticized CBNRM as a mistake which had
failed to prevent loss of biodiversity, and argued for a move away
from CBNRM and a return to parks with 'hard barriers' which
would be imposed by military-style force. His arguments were
echoed by a second high-profile book published in the same year by
John Frederick Oates, *Myth and Reality in the Rainforest*. Oates's book
also criticized community-based approaches and argued strongly in
favour of hard parks with military enforcement.[25]

Coercive anti-poaching methods rely on groups of armed and
trained rangers engaged in military-style operations. Since
commercial poaching gangs are also often heavily armed, rangers
can find themselves regularly engaged in shoot-outs. In effect, if we
support coercive forms of anti-poaching we have to be prepared to
accept that rangers and suspected poachers will be injured or even
killed in confrontations over valuable wildlife. However, it is
important to consider whether these approaches to conservation
produce serious social injustices, and if they do then what impact
do they have on conservation. In order to answer this we need to
examine the range of anti-poaching practices. What is quite clear
is that both military-style anti-poaching and community-based
approaches only really attempt to tackle hunters operating in
parks. These hunters are only one small part of a much bigger
system which is not dealt with effectively via anti-poaching patrols
or incentive schemes for local communities. In short, current anti-
poaching strategies do not have the capacity to deal with the
groups that organize the poaching at a higher level to feed the
demands of markets in ivory, rhino horn and tiger bone.

The international commitment to, and public acceptance of, a
war on poachers is related to the powerful international profile of
key figures in conservation, two of the most famous of whom are

Dr Richard Leakey and Dian Fossey. Leakey and Fossey are central to Western imaginings of conservation. They are portrayed as saviours of our most cherished wildlife, especially elephants and gorillas, and as people who have faced danger and made sacrifices to protect wildlife. Their high public profile is key to understanding how they influence public opinion.

Dr Richard Leakey, son of internationally famous archaeologists Louis and Mary Leakey, is one of the most famous figures in conservation. His book *Wildlife Wars: My Battle to Save Kenya's Elephants* details how he was always in favour of military-style approaches to anti-poaching, and his current conservation venture, the NGO Wildlife Direct, continues to reflect this attitude. In the 1980s, Leakey claimed he was waging a 'war' to protect wildlife. He made it clear that the rangers in the Kenya Wildlife Service were facing heavily armed and organized poaching gangs. After the announcement that Kenya was to pursue shoot-to-kill policies in 1988, he gleefully declared that soon journalists would be able to take photos of dead poachers instead of dead elephants. Large-scale poaching in East Africa in the 1980s galvanized international conservation NGOs to start campaigns against the ivory trade and call for effective anti-poaching measures. When Leakey was Director of the Kenya Wildlife Service (KWS), critics pointed to its increasing militarization under his leadership. Leakey has persistently developed and argued in favour of military techniques to deal with poaching. He states: 'given the high commercial value of ivory, I knew this battle was going to be part of a costly, drawn out war, one fought on many fronts: poachers, middle men and corrupt officials'.[26]

Leakey was concerned about the high levels of corruption in the KWS which facilitated poaching and smuggling; in *Wildlife Wars* he also mentions the rumours about the involvement of family members of former President Kenyatta, especially his wife, but dismisses them as merely speculation. He pointed to the involvement of foreign diplomats who used the confidentiality privileges of the 'diplomatic bag' to move ivory out of East Africa and into Asian markets.[27] Government authorities, in conjunction with international NGOs, have placed the blame on poaching gangs that hail from neighbouring states. This was certainly the case during the 'wildlife wars' of the 1980s in Kenya, when the government blamed Somali *shifta* (bandits) for crossing the border to wipe out the elephant population.[28] Despite his apparent awareness of the complexities of the poaching wars of the 1980s in Kenya, Leakey's main 'weapon' in the war was to be armed rangers engaged in violent anti-poaching campaigns which targeted poaching gangs. But this approach tackled individual poaching gangs which were only one manifestation of a wider problem: organized poaching which was the result of global dynamics of demand for ivory.

Similar issues are raised by the career of Dian Fossey, whose life and work with gorillas in the Virungas mountains in Rwanda was detailed in her book *Gorillas in the Mist*, which was then dramatized as a Hollywood film starring Sigourney Weaver as Dian.[29] The film portrays a woman who was committed to caring for and protecting 'her' gorillas against all odds, and Fossey's public profile has been raised further by numerous books about her, including *No One Loved Gorillas More* by Camilla de la Bedoyere.[30] The romantic idea of a courageous woman risking her life to save gentle giants has remained a key feature of Fossey's public image. The mysterious circumstances surrounding her death in 1985 at the Mount Karisoke Research Centre in Rwanda have also served to keep Fossey and her work in the public eye. She was killed with a blow to the head by a machete, and speculation continues about who murdered her. Theories include poachers, her own staff, or hired assassins who worked for the tourism interests that Fossey was obstructing. Romaticized portrayals have had a significant impact on public perceptions of Dian Fossey and of the problems facing mountain gorillas in central Africa. But her story provides us with an example of the complicated knot of ethical dilemmas involved in anti-poaching strategies.

To this day many of the gorilla-watching trips in the Parc National des Volcans in Rwanda incorporate a trip to see the Mount Karisoke Research Centre where Fossey worked, and where she and Digit, one of the most famous of 'Fossey's gorillas', are buried. Her descriptions of the Virungas areas are used by the Rwandan Tourism Authority to encourage visitors to the country – despite the fact that Fossey was vehemently opposed to the habituation of gorillas for tourism, and did her utmost to frustrate all attempts at developing gorilla tracking in Rwanda.[31] But her appeal is immense and is central to public perceptions of gorilla poaching and the role of conservationists in protecting wildlife.

Supporters of Fossey and her work are reluctant to take on board criticisms levelled at her interactions with Rwandan parks staff, local villagers, poachers and fellow researchers. Harold T. D. Hayes's biographical work *The Dark Romance of Dian Fossey* is very critical of Fossey and her methods.[32] Her supporters have argued that the book and its criticisms are based on dubious evidence, yet other researchers who worked with Fossey at Mount Karisoke have also pointed out that while she could be fantastically captivating and was genuinely committed to gorilla conservation, she could also be very stubborn, cruel and petty. For example, in *In the Kingdom of Gorillas*, Bill Weber and Amy Veder write of their conflicting feelings about working with Dian Fossey.[33] Their book details their time working at Mount Karisoke with Fossey and recounts stories of how she used to beat suspected poachers with barbed wire and elephant nettles, and how one researcher left the camp and refused to work with her again after she played out a 'mock execution' on a suspected poacher. They also singled out her attitude towards Rwandan parks staff as particularly negative and racist. Yet Fossey is still treated uncritically by the conservation movement for her dedication to saving gorillas, while the more unpalatable aspects of her methods tend to be glossed over. However, her confrontational and often violent tactics ensured she was deeply unpopular with the communities that surrounded the Mount Karisoke Research Center. These methods raise the questions of whether mock executions and torture are either morally defensible when used in pursuit of wildlife conservation or effective in the longer term. After all, confrontation and coercion rarely persuade people to support your cause.

One of the key terms in this story is '*suspected* poacher'. In the drive to save wildlife we often overlook human rights. When suspected poachers are injured and killed in confrontations with rangers over wildlife, in effect they have been tried, convicted and put to death as a result of their unwelcome presence in national parks. The case of the 'rhino wars' of the Zambezi Valley in Zimbabwe is instructive here. In 1986 President Robert Mugabe approved the introduction of a shoot-to-kill policy. In an address to the House of Assembly in 1987, he redefined clashes between poachers and rangers as 'contacts', which meant that poachers were then treated in the same way as RENAMO rebels from Mozambique. This legitimized the full deployment of security forces, who were given the right to shoot on sight. In addition, rangers were protected from prosecution for injuries and deaths caused by anti-poaching operations under the 1989 Wildlife Indemnity Act, so in effect they could act with impunity.[34] In 1990 the Zimbabwe government and wildlife NGOs were pleased to announce that they had successfully protected wildlife because more poachers were killed than black rhinos; between

1984 and 1993 approximately 170 poachers lost their lives; nearly all were Zambians except for a few Mozambicans and Zimbabweans. These celebratory statements continued despite protests from local communities that the Parks Department and Zimbabwean Army were using aerial firepower against innocent people; representatives of the villagers in the area claimed that many of the people shot and killed from helicopter gunships were merely using the protected areas and rhino habitats as shortcuts between villages and that they were not engaged in any kind of criminal activity.

For conservation agencies, the use of shoot-to-kill policies is justified as being for the benefit of critically endangered wildlife; and they claim that when rangers are faced with groups of armed poachers they must be able to respond. At its most basic: the ends justify the means. This raises ethical questions about the rights of people who are only suspected of a crime, balanced against the rights of wildlife and the needs of conservation. Furthermore, even if the poachers were found to be guilty through an acceptable legal process, the question remains of whether their crime should be punished by the death penalty.

As we have seen, Richard Leakey's current venture is Wildlife Direct, which works in African countries including the DRC and Kenya. Wildlife Direct is an interesting development in the conservation sector. Its website is filled with blogs from parks staff and anti-poaching units; their online diaries describe the daily problems of encountering poachers, of finding dead elephants and of the need for donations to support their work. Wildlife Direct also uses text messaging so that members can update the organization immediately when a poacher is captured or shot, or when rangers encounter rebel fighting in eastern DRC. In the donations section of the website, Wildlife Direct make clear their commitment to military-style enforcement, while assuring potential donors that their money is going directly to a good cause:

> Wildlife Direct was founded and chaired by Dr Richard Leakey to help bridge this gap by using the internet, and especially blogging, to connect and empower like minded people . . . We believe in good conservation that can be trusted, and so when you donate money to help pay for the rations of rangers on patrol in Congo, we want you to know that this is a good use of money and that it is exactly where your money has gone.[35]

Potential donors can choose to pay for clothing, equipment or rations for rangers, so they can purchase everything from boots to bullets. This directly links the donor with the anti-poaching and ranger patrols in various protected areas around the world. But this may not necessarily be an overwhelmingly positive step for conservation. Members of the public and corporate donors trust that the rangers are 'doing the right thing': they are securing important species and key habitats, and so their supporters tend to overlook the actual practices involved in achieving those aims. What we see in the conservation sector is that questions of ethics, rights and justice often get overlooked in the drive to deliver wildlife conservation which is thought of as 'an obviously good thing'. The presentation of military-style protection of wildlife as a *good* thing to do silences the debate about whether it is the *right* thing to do, and even whether it is an *effective* thing to do.

This drive to do a 'good thing' for conservation means that NGOs that otherwise would not be seen to be supporting violence have turned a blind eye to it. In turn, NGOs have justified their position on the basis of protecting the world's last remaining pockets of biodiversity for future generations. This is assisted by a simplistic definition of rangers as 'wildlife protectors' and poachers as 'wildlife destroyers'. One of the most controversial donations by an NGO came from WWF-International, which donated a helicopter for the Zimbabwean anti-poaching effort in 1987. The helicopter was used in the shoot-to-kill policy, and it was trumpeted in a Zimbabwean newspaper as the helicopter used to shoot a poacher in the Sapi safari area.[36] However, the helicopter quickly turned into a public relations disaster as WWF-International became embroiled in row over human rights. The provision of the helicopter was severely criticized by development and human-rights organizations for its role and obvious support for shoot-to-kill policy. WWF-International, whose stated policy is not to provide weaponry or funding for weaponry, denied it was ever intended to be used as a helicopter gunship, but Raymond Bonner argues that there was fierce internal debate about its use as a gunship.[37] The US$6000 of funding needed to run the helicopter was eventually withdrawn. Publicly, WWF-International stated that their decision was due to lack of funds and because they did not agree with Zimbabwe's conservation policy, specifically culling.[38] However, it was suggested that the helicopter was withdrawn because of international criticism of WWF's involvement in shoot-to-kill methods. From this case it was clear that although NGOs could use their association with anti-poaching to raise their own profile in a positive manner, it could also be disastrous for their international standing.

NGOs and national governments have gone one step further in recent years, engaging the private sector to enforce conservation rules. As a result, protecting wildlife has been used to justify the hiring of mercenary forces for anti-poaching patrols. Following the end of the Cold War and the end of apartheid in South Africa, former military personnel began to move into private sector security. In South Africa, former soldiers from the apartheid era South African Defence Force carved out a new niche in conservation. In many ways the skills needed by rangers are similar to those possessed by military personnel, including survival skills, knowledge of weaponry and the ability to plan operations. Therefore from the mid-1990s the conservation sector increasingly saw the use of private military companies to enforce protected areas. This may seem outlandish but there is a scheme by African Rainforest and River Conservation (ARRC), operating in the Central African Republic, whereby the sale of diamonds dug up in their area are used to fund their privately owned reserve. To secure biodiversity within the area the ARRC has hired mercenaries who previously served in the Rhodesian army and the SADF during the apartheid era. These private soldiers were hired to organize anti-poaching patrols to tackle poachers who the ARRC say are operating across the border from Sudan. These may seem like extreme measures, but they are not uncommon in wildlife conservation in Africa where violent methods can be used under questionable authority, leading to serious human-rights abuses. International NGOs may engage in projects for the highest moral reasons and genuine commitment to wildlife conservation, but through their logic of responding to the situation on the ground in a pragmatic way their actions can be more like those of companies engaged in diamond and oil extraction.[39]

In South Africa the many privately owned wildlife reserves, especially those bordering Kruger National Park, have also used private security companies to protect their wildlife stock; after all, the wildlife contained in the private reserves is the primary tourist attraction, and therefore they are financially valuable. In the case of these privately owned reserves protected by private armies we need to be clear why the wildlife is being protected, who it is being protected from and what purposes this serves.

3.2 Beach fenced off for tourism, Lake Victoria, Uganda (2007)

Often it is not to protect critically endangered species for future generations to enjoy, but to secure valuable property generating income from tourists.

The argument that tourism will make protected areas financially viable has become omnipresent and is promoted by conservation NGOs, national governments, private companies and international donors alike (as discussed in chapter six). But we need to be careful that we are not supporting a process whereby some parts of the world become heavily fortified enclaves where wealthy tourists can enjoy themselves to the detriment of the human communities that surround them. This results in a pattern where the use of deadly force is regarded as necessary (or even desirable) to defend landscapes and wildlife populations that are deemed useful or saleable on the international market. This model of carving out territories for wildlife ignores the fact that such enclaves and biodiversity hotspots are reliant on the communities, landscapes and animal populations that surround them and link them together, so there is little point in fortifying tiny islands of biodiversity.

CONCLUSION

Poaching is multifaceted. The simplistic definition of poaching as all hunting not sanctioned by the state utterly fails to capture the vast range of practices in the spectrum between subsistence and commercial poaching. The question of where we draw the line between hunting and poaching is fraught with ethical and practical difficulties. However, in the international conservation movement there is a clear turn towards the use of blanket bans and the use of violence and coercive methods of enforcement. This means that picking up a dead honey badger in a rural area is caught in the same legal net as commercial ivory poaching: both are banned. For communities that rely on small-scale subsistence hunting, the enforcement of bans has serious negative effects, and the removal of historical rights to use wildlife to meet everyday needs is regarded as an injustice by those communities.

Local communities react by engaging in everyday acts of resistance: cutting fences, hunting wildlife, setting traps and snares, engaging in non-co-operation with conservation agencies. For example, Jacoby notes that in New York State's Adirondack Mountains, state officials were frequently frustrated by local resistance, which ranged from surreptitious violations of game laws to arson and even murder, and in many of the national forests created at the start of the twentieth century, squatting, timber stealing, and illegal hunting were commonplace.[40] Even today, conservation is replete with complaints from NGOs and government parks agencies about encroachment in various forms: setting wire snares to catch game, grazing livestock within park boundaries, cutting firewood and vandalism of park infrastructure. These are the only weapons available to them, and although they can be effective in terms of frustrating the conservation objectives of NGOs and state authorities, they are no match for the superior fire–power that rangers are often equipped with. Furthermore, rangers are often indemnified (legally protected) against prosecution for crimes committed while engaging in anti-poaching patrols. In contrast, local communities find their activities are criminalized and they are punished for them, sometimes with deadly force. The redefinition of hunting as a criminal act of poaching also demonstrates how conservation is not just a victim of illegal practices, but is itself involved in producing and sustaining the 'illicit'. But in the longer term, resentments over the imposition and enforcement of such rules mean that conservation agencies risk alienating the very people who live with wildlife, and who may well hold the key to securing wildlife, populations for future generations to enjoy.

The ways in which we think about and define poaching also sets the acceptable limits of anti-poaching operations. If we define hunting as a crime against nature and future generations, coercive methods to prevent wildlife loss start to appear more acceptable. The use of violence against suspected poachers can mean that parks agencies engage in human-rights abuses, and we need to consider whether shoot-to-kill policies actually mean that people are being subjected to extra-judicial executions. These ethical dilemmas in protecting wildlife are compounded by the arrival of private armies used to enforce protected areas for governments, NGOs and private philanthropists hoping to 'save' wildlife. It is not clear who gives authority to private companies to injure or kill people suspected of poaching. Furthermore, it is unclear where accountability lies: governments, international conservation organizations or the private individuals who hire them to protect the expanding number of privately owned wildlife estates.

Anti-poaching measures sharply demonstrate how the social injustices inherent in many conservation policies mean that conservation initiatives that start with good intentions can end up being highly unjust and ultimately counterproductive. International tourism has made some species and habitats financially valuable, providing a powerful rationale for use of coercive and violent methods of protecting wildlife. Communities find themselves branded as unwelcome trespassers in national parks where wildlife serve not only as an important part of an ecosystem, but also as a significant tourist attraction. The assumption that the ends justify the means results in a situation where the international conservation movement and their supporters around the world assume they are making ethical and environmentally sound decisions to save wildlife, but are in fact supporting practices that have counterproductive, unethical and highly unjust outcomes.

CHAPTER FOUR

RHINO HORN, IVORY AND THE TRADE BAN CONTROVERSY

1989 WAS A YEAR OF momentous change: the world was fundamentally reorganized as the Soviet Union collapsed and the traditional Cold War divides of East versus West suddenly disappeared. 1989 was also a year of momentous change in conservation: it was the year of the ivory ban, one of the most important and controversial decisions of the twentieth century. The decision to ban the ivory trade was taken at the international level, but it is one of the best examples of how conservation rules can suddenly change the ways people can manage and use wildlife. The ban meant that a whole global industry was shut down almost overnight. But since 1989 there have been persistent campaigns to reopen a legal ivory trade. Controversy rages on over whether trade bans can really provide a long-term answer for conservation. The bans on rhino horn and ivory trading have caused deep divisions.

Bans appear as a simple and effective solution, the obvious response to poaching and trading, but the problem is that they are a 'one-size-fits-all' solution. They fail to take account of the differences in rates of poaching, populations in range states, management regimes and conservation policies in different countries. Bans have also tapped into inequalities between the richer world and the poorer world. A number of African countries resent the powers of Western NGOs and states, especially over decisions about ivory and rhino-horn trading. Elephants and rhinos are two charismatic species – they are keystones of the wildlife safari industry and icons of the conservation movement. They are also both subject to high-profile campaigns to maintain an international trade ban.[1] This chapter will explain the controversies that continue to rage over the ivory and rhino-horn trade. It will look at the background to bans, and the continuing struggles over the rhino-horn and ivory trades, including the arguments in favour of reopening the ivory trade to *achieve* conservation.

RHINOS

Imagine a typical farming scene. The livestock are humanely reared on the farm, allowed to run around outside, and they produce a very valuable product which can be quickly removed without hurting or killing the animal. This is the case with sheep used for wool production, which most of us have no ethical problems with. The sheep are well cared for, shearing them does not hurt them and they quickly grow a new coat. But what if the livestock were something more exotic, like a rhino? Rhino horn can be removed without hurting or killing the animal, and rhinos quickly grow a new horn. This has led some conservationists to suggest that the best interests of rhinos may be served by starting up legal rhino-horn farms. For them, there is no ethical difference between the sheep and the rhino, as long as the rhinos are bred in captivity and domesticated, rather than being caught in the wild. This view suggests that it is hypocritical to use wool from sheep but then object to rhino farms. Of course there are strong objections to this argument, the strongest of which is the potential relationship between farmed and illegally poached rhino horn.

The trade in rhino horn is undoubtedly one of the hottest conservation topics. It is an important moral theme for many members of CITES and for wildlife conservation NGOs. The demise of rhino populations around the world is a damning indictment of the failings of conservation. The trade in rhino horn was banned in 1977 but rhinos were decimated through the 1970s and 1980s. The figures are hard to argue with: between 1970 and 1992, the black rhino population decreased by 96 per cent according to the International Rhino Foundation. In 1970, it was estimated that there were approximately 65,000 black rhinos in Africa, but by 1993, there were only 2,300 surviving in the wild. However, it is not all bad news, and the picture is mixed. Southern white rhino populations have experienced the opposite trend. At the turn of the twentieth century there were only 50–200 left as a result of hunting and poaching, but numbers have recovered to approximately 17,500 today. Similarly the Indian one-horned rhinoceros has recovered in the past 100 years, from fewer than 200 in the early twentieth century to over 2,000 now.[2]

It is difficult to estimate how many rhinos are left in the world but Save the Rhino International suggested that there were 4,240 black rhinos and 17,500 white rhinos in Africa in 2009. They identify populations in most African states as either up or stable when compared with figures from December 2005. In Asia, some species of rhino are highly endangered, with 2,620 Indian rhinos, 200 Sumatran and just 50 Javan rhinos left in the wild; it is estimated that the only Asian rhino population that has increased is the Indian one-horned rhino – and that is a rough guess.[3] These figures disrupt the message that poachers and illegal traders are devastating rhino populations. These recoveries are important conservation success stories; but concerned NGOs point out that the rise in rhino populations may be rather precarious.

These arguments form a critical part of the debates about whether the rhino horn trade ban has been a success or not. The rhino-horn trade was one of the critical issues in the foundation of CITES, and the international trade in rhino parts was banned in 1977. However, the illegal trade continues and is seen as evidence that CITES is failing and that trade bans simply do not work. But the problems with CITES also show us how conservation rules can create and sustain criminality. Trade bans reduce the supply of wildlife products and this can make them even more valuable. One of the main criticisms of the trade ban is that it made rhino horn even more lucrative and therefore helps to sustain the illegal trade. I will explore this in greater detail later, but first it is useful to discuss where the continuing demand for rhino horn comes from.

The Rhino Horn Trade: From Aphrodisiac to Aspirin

It is tempting to simply repeat the urban myth that rhino horn is used as an aphrodisiac. The aphrodisiac myth informs many of the stereotypical criticisms made of Chinese medicine: that it is responsible for wasteful use of wildlife and that it has no real medicinal value. However, studies by rhino specialists such as Esmond and Chrysee Bradley Martin reveal that such stereotypes fail to capture the complex historical uses of rhino products, and the cultural attachments to them. It may well be that the myth began with traders, conservationists and travel writers who bought rhino horn from the Indian traders they encountered in

East Africa. Europeans wanted to know why the horn was in such demand in China, and it is thought that the Indian and African traders claimed it was an aphrodisiac with an eye to increasing demand for rhino horn in Europe.[4]

As with the rest of the wildlife trade, demand for rhino horn is driven by wealthier communities around the world, not by poverty. There are two main sources of demand for rhino horn: traditional ornamental dagger handles, or *jambiya*, in Yemen which are carved from rhino horn, and the use of powdered rhino horn in Traditional Chinese Medicine (TCM); here I will concentrate on the role of TCM as the main source of sustained worldwide demand. East Asia has historically been the main region of the world where rhino horn is consumed as an ingredient in medicines and used for decorative carvings. Sometimes these roles have been fused together. For example, cups made from carved rhino horn were used in Chinese royal households to determine if a drink was poisoned: it was reputed that the acidity of the poison would react with the alkalinity of the rhino horn cup and cause the drink to 'fizz'. However, since the later twentieth century rhino horn has been prized mostly for its importance to TCM.[5] This demand developed and expanded partly as a result of Mao Zedong's promotion of traditional medicines after the Second World War, and more recently owing to rising incomes within China. Although the domestic trade in rhino horn was banned in China in 1993, the illegal trade has continued since then.[6]

While China and Taiwan are the main consumers of rhino horn, the demand for rhino horn-based products extends outside these countries to Chinese communities around the world. Again it is tempting to repeat the usual assumptions that the use of rhino horn has no real medicinal value and is a wasteful use of valuable, important and much loved wildlife. But in order to understand why the trade continues, and even how it might be shut down, we also need to know why there is such an attachment to the use of rhino horn in TCM. The properties of rhino horn have been denied by Western medicine, and conservation organizations have claimed that consuming medicine that contains rhino horn is no different from biting your fingernails. However, more recently, it has been accepted that rhino horn has the same properties as aspirin, which is used to treat common ailments such as headaches and fevers. In the text on which much of Chinese medicine is based, the *Pen Ts'ao Kang Mu*, its main uses are cited as for reducing fever, for heart complaints and other preventive purposes.[7] This is similar to the use of aspirin in Western medicine to bring down fevers and as a blood thinner for heart patients.

The demand for rhino horn resulted in a rise in poaching. However, poaching rates are also linked to changing dynamics in world politics. The rise in demand and increase in rhino-horn prices was closely related to illegal trading syndicates that link East Asia and Africa. As with the cases of caviar, tiger products and abalone (covered in chapter one), the involvement of powerful and influential criminal cartels such as the Mafia, Triads and Yakuza presents a real problem for enforcement of international regulations like the CITES trade ban. Criminal networks who deal in drugs and people trafficking were attracted by the wildlife trade because wildlife crime carries relatively low penalties when compared with drug trafficking and the chances of being detected are much lower because wildlife crime is not a major priority for law-enforcement agencies. Finally, the potential profits compare very favourably with profits to be made from trading in other illegal goods.[8]

Taiwanese dealers played a significant role in keeping the trade going in the 1990s. Rhino horn commanded such high prices that dealers viewed buying horn as an investment, and the EIA reported in the early 1990s that 10 tonnes of horn were stockpiled in China and Taiwan.[9] It appeared that dealers were trying to cash in on the high prices for rhino horn by making it 'scarcer' through stockpiling. Taiwan has presented a real problem for international enforcement of wildlife legislation. It is a type of black hole for international legislation because of its special position in the international system, which arises from its history and relationship with China: it is not recognized by the UN, and so cannot be an independent member of CITES. Taiwan banned the import and trade in rhino and tiger products in 1985, but illegal trading continued. For example, in 1993 Taiwan held some of the largest stockpiles of rhino horn in the world, built up prior to the international CITES ban in 1977 and before trade controls were tightened in Taiwan, Hong Kong and Macao in 1985. The Taiwanese government refused to destroy these stockpiles in 1993.[10] The continued illegal trade in rhino horn has prompted other states where rhino horn may be consumed as part of TCM to develop legislation to deal with the problem. For example, in 1994 the US passed the Rhinoceros and Tiger Conservation Act which banned the sale of any products which listed rhino or tiger parts as an ingredient.[11]

Are Bans the Solution?

The debate over whether the rhino horn trade ban has worked or not rages on. Broadly there are two camps: one side argues that a trade ban is the only way to save rhinos, and that current failings are related to problems with enforcement rather than with the ban itself. The counter argument is that the ban has demonstrably failed, as indicated by the continuing illegal trade and declining rhino populations around the world, so the only way to save rhinos is to legalize the trade, control it and use the funds generated from it to enhance species conservation. The debates mirror arguments over the drugs trade: do bans merely drive the trade underground and make it harder to regulate? Would it be more effective to legalize and control the trade? Would it be beneficial for conservation to allow the state to trade in lucrative wildlife or to tax the trade as a means of raising much-needed revenue? Either way, the figures on the continuing illegal trade in rhino horn and declining rhino populations are hard to dispute.

It is difficult to obtain accurate prices for rhino horn given that the trade is illegal, but it is a valuable commodity: IUCN estimated that the price of rhino horn experienced a twenty-one-fold increase between 1975 and 1980 alone.[12] The high prices paid and the persistent demand mean that rhinos are a target for poachers and traders. However, these same two factors, profits and demand, also explain the disagreements in CITES over whether lifting the trade ban would in fact help to save the rhino. Those in favour of lifting it argue that allowing governments, private business and local communities to derive full economic value from rhinos would produce incentives to conserve the species. In the mid-1990s, the Zimbabwean Department of National Parks and Wildlife Management argued that 'farming' rhinos for their horn might present a possible solution. The Parks Department had started a policy of dehorning rhinos to make them less attractive to poachers. In the course of the dehorning programme they discovered that the horn could be removed painlessly and that it grew back at the rate of approximately one inch per year. This raised the prospect of legally farming rhinos for their horn, which would produce a sustainable and renewable crop. However, there was no way to develop the idea because of the trade ban. It also became clear that poachers still killed dehorned rhinos for the valuable stump, or so that they did not waste time tracking the same rhino again.[13]

The case of rhino horn farming reminds us that international legislation can strictly define and delimit what can and cannot be done in the conservation arena. The international legal framework of CITES essentially outlaws some of the avenues that the conservation sector might want to try out. Of course there are important debates to be had about animal welfare and whether this is an appropriate and ethical way to treat rhinos. Numerous animal welfare-oriented NGOs such as EIA and IFAW were opposed to the scheme on moral grounds; they also argued that any legal trade in farmed horn would encourage the illegal trade and prompt another devastating round of poaching.

The international ban on the rhino-horn trade has limited the possible solutions for rhino conservation, and frustrated attempts by Southern African countries to use the trade as a conservation tool. Although rhino-horn farming did not develop as an industry, another form of 'consumptive use' of rhinos has developed in South Africa and Namibia.[14] The debates on the effectiveness of the rhino-horn trade ban have been complicated by the case of rising white rhino populations in South Africa, where sport hunting of white rhino has been allowed since 1968. Richard Emslie, the Scientific Officer for the IUCN Species Survival Commission (IUCN-SSC) African Rhino Specialist Group states that since hunting began, numbers of southern white rhino have increased from 1,800 to 11,100 in the wild, with a further 740 in captivity worldwide. Allowing hunting has helped give white rhinos an economic value, which has encouraged conservation by the private sector and by local communities. For example, by 2003, 3,250 of Africa's southern white rhino were privately owned. Rhinos in South Africa have a high economic value: live males auctioned in KwaZulu-Natal for non-hunting tourism initiatives fetched an average price of US $21,130, while it has been estimated that a black rhino trophy hunt might fetch about US $200,000, almost ten times the live price.[15]

The Professional Hunters Association of South Africa (PHASA) argues that the revenues generated from sport hunting have allowed white rhino populations to grow on private reserves and ranches. Here we can see that the value of rhinos and the ability to export their horns as trophies has directly contributed to the conservation of a rare species. Supporters of this approach are keen to argue that this is much more successful than the CITES trade ban. The success with white rhinos led to a decision by the thirteenth CITES Conference in Bangkok (2004) that the numbers of the more endangered black rhino had increased to a sufficient level to allow South Africa and Namibia to hunt five black rhino per year. In the case of Namibia, the Ministry of Environment and Tourism argued that their management had resulted in a more than 1,000 per cent increase in black rhino populations since 1960, and that hunting five older males each year was entirely sustainable. However, in recent years there have been some concerns in South Africa that the sport hunting permits were being misused to export illegally hunted rhino horn. In 2008 the Environmental Affairs Minister, Marthinus van Schalkwyk, suspended the export of white rhino trophies from South Africa because the Ministry had discovered that permits for sport hunting trophies were being used to export illegally poached rhino horn. Permits were issued for hunts that never took place and then the permits could be used to move illegally hunted rhino horn out of the country.[16]

These concerns have been echoed by debates within CITES itself. Kenya issued a counterproposal at the fourteenth CITES Conference in the Hague in 2007 to prevent South Africa and

Namibia from selling their five hunting permits for black rhinos. The proposal was defeated, to the satisfaction of the South African representative at CITES: the permits are worth up to US$1 million per year. A further Kenyan proposal, that all rhino-horn stockpiles should be destroyed, was also defeated. However, CITES did agree on greater co-operation in monitoring and enforcement of the rhino horn trade ban among African states.[17]

The recovery of white rhino populations in South Africa, and the mixed picture of rhino recovery around the world do raise some important questions for whether a ban remains the best mechanism for conserving rhinos in the longer term. These divisions are even deeper over the ivory trade. There are bitter clashes at each CITES meeting which reveal why trade bans do not always provide a long-term solution for conservation.

THE HISTORY OF THE IVORY TRADE

African elephant ivory is particularly sought after because the tusks are larger and easier to carve than ivory from Asian elephants. The tusks are in fact large teeth which keep growing throughout the life of the elephant. They can grow to impressive sizes; the American Museum of Natural History in New York displays two tusks of 335 and 349 cm respectively, while the Natural History Museum in London displays two tusks weighing 102 and 107 kgs.[18]

The ivory trade has a very long history, and demand for ivory has shaped trading routes, built civilizations and underpinned empires, especially in Africa. For example, the ancient Egyptians referred to a mythical land (probably somewhere on the coast of the Horn of Africa) called 'the Land of Punt' which was the source of gold, slaves and ivory. In the nineteenth and early twentieth centuries ivory was much in demand in Europe and in North America, where it was carved into jewellery, ornaments,

4.1 Ivory ornaments at a Buddhist temple in Sri Lanka (2003)

billiard balls and piano keys.[19] This means that the cultural and historical attachment to ivory was not as strong as in Asia; in the Western world ivory was associated with trinkets and leisure pursuits. But it was still an important commodity that had a major role to play in international trade. For example, in the late nineteenth century, ivory helped build and expand European empires around the world. Indeed, British imperial expansion in Africa was partly subsidized by hunting for ivory; as we saw in chapter three, expeditions by European hunters keen to exploit healthy game populations were important precursors to formal colonization.[20] Much like the illegal ivory trade of the late twentieth century, these historical trading routes crossed international borders and ivory was sold through both legal and illegal outlets. So the poaching wars and the challenge of distinguishing between legal and illegal ivory that conservationists faced in the 1980s were not necessarily new problems.

There is a long history of ornamental ivory carving in East Asia; the carvings range from small statues to large-scale ornaments that depict landscapes and buildings. There is a good example of this in the United Nations building in New York: in the hallway there is an enormous ivory carving which was a gift to the UN from the People's Republic of China. But there is also a cultural attachment to ivory in Japan, where it is used to make traditional musical instruments and *hanko*, or personal seals, which are shaped rather like a fat piece of chalk with an individual design carved into one end. The seal is used to print this unique 'signature' on correspondence and agreements. Most *hanko* seals were made from wood and tipped with ivory until the economic boom in Japan after the end of the Second World War. Unlike other wildlife which is used for medicine, food and clothing, ivory is primarily used for ornamental purposes. In theory it should be easier to campaign against the ivory trade because it is not needed to sustain human life.

4.2 Ivory carving at the UN headquarters in New York, a gift from the People's Republic of China (2006)

PAVING THE WAY FOR THE IVORY BAN

In the late 1980s poaching was devastating elephant populations in some parts of Africa. The global media was filled with gruesome footage of dead elephants in Kenya and Tanzania. The ivory trade was deemed to be responsible for the massive decline in elephants – numbers crashed from 1.3 million to just 600,000 during the 1980s; Kenya had 65,000 elephants in 1979, but only 17,000 were left in 1989.[21] The evidence seemed clear: Africa had witnessed a massive decline in elephant populations, caused by cruel and greedy poachers hunting to feed the insatiable demand for ivory. It looked as if the world was sitting idly by while human greed drove elephants to the brink of extinction.

It is difficult to say exactly where the ivory campaign came from. In the 1970s and 1980s there were discussions about the ivory trade in CITES. The critical moment turned out to be 1988, when the US-based African Wildlife Foundation (AWF) declared it the Year of the Elephant. But this did not result in an instant backing for a full ivory ban; that did not come until late 1989. Conservation NGOs played a vital role in this campaign. The ivory trade ban debate appeared on the international stage just as NGOs were gaining greater levels of power in the international system. Conservation NGOs were brilliant communicators. Since they were international organizations they were able to say they were not tied to any particular national interests, and they truly had the interests of elephants at heart. This is a very convincing line of argument – as discussed in chapter two, the NGOs do not represent countries, they develop their own research and information on conservation, they design and run impassioned campaigns for wildlife, and are able to present themselves as morally driven actors.

Like many other people in 1989, I proudly wore a WWF t-shirt with the slogan 'Ivory Looks Best on Elephants' emblazoned across it; it was one of my favourites, along with my anti-whaling t-shirt from Greenpeace. WWF appealed to the idea that our love of trinkets and fashion was killing these wonderful, caring and sensitive creatures halfway across the world. The sloganeering was simple, memorable and extremely effective. At the time one of the world's biggest conservation organizations, WWF realized that its response to the ivory issue could have a major effect on its future as its memberships and the wider public looked to it to provide leadership of the ivory debate.

The African Wildlife Foundation also played a vital role in pushing forward the campaign for a total trade ban. As part of their 1988 'Year of the Elephant' initiative, they hired an advertising firm to work on the wording and imagery used in an 'Urgent Memorandum', part of a direct mailing campaign aimed at educating the American public about the impact of buying ivory trinkets and jewellery. In the related press conference, Diana McMeekin of AWF urged the American public not to buy ivory, but still did not call for a total ban. The AWF had a long-term conservation interest in Africa and was trying to be sensitive to alternative ways of seeing and living with elephants.[22] Still, the purpose of the press conference was to shock the public to the extent that they stopped buying ivory, and it was hoped that the conference would also raise funds for AWF.

AWF's publicity had a life of its own once it was out in the public arena, and it caught the attention of a number of animal-rights organizations. One of the key players in creating an atmosphere of crisis in the run-up to the ivory ban was the UK-based EIA. Established by Allan Thornton in 1985, its focus was campaigning rather than fundraising or financing conservation projects. Thornton had previous experience as a member of Greenpeace and had engaged in their direct action campaigns over seal culling in Canada. As a small organization with a limited membership base EIA needed to raise funds for its ivory campaign, and the money was provided by Christine Stevens of the US-based Animal Welfare Institute. Supporters of EIA argued that they were motivated by a real desire to save elephants; critics accused them of engaging in cynical media manipulation to raise their own profile. Either way, EIA produced a highly influential report entitled *A System of Extinction: The African Elephant Disaster*, which was circulated to all CITES representatives. It denounced the pro-ivory trade position, accused CITES of presiding over the extinction of the elephant and argued that Southern Africa's continued support for a legal ivory trade was determined by the need to protect commercial interests in the culling, hunting and ivory industries.[23]

EIA were also brilliant at producing film footage for use by prime-time news channels, which served to shape public opinion on the ivory trade. The films were instrumental in shocking the public in Europe and North America about the scale of elephant slaughter in East Africa and meant it was very difficult for any Western governments to vote to continue the ivory trade at CITES meetings. Finally, EIA was very skilled at building alliances within CITES. It teamed up with WWF-US to draft a proposal to ban the ivory trade; under CITES rules they needed a representative of a national government to propose it at the conference. Since Kenya's conservation record was too tarnished by accusations of corruption, they persuaded Tanzania to table the proposal at the 1989 meeting.[24]

Dissenting Views

The problem was that while we all thought we were saving elephants at the time, the headlines and NGOs campaigns over-simplified the issue. In fact, not all conservationists, and certainly not all African states, were in favour of a ban. Their side of the argument did not make it on to global news reports or into glossy wildlife magazines. It was a far from glamorous story about human–wildlife interactions, rural poverty and divisions over wildlife management options.

The problem was (and still is) that large numbers of elephants are found outside national parks, and so they inevitably have to share space with people. Our image of elephants is of herds gently moving through the wilderness, but in fact elephants move in and out of national parks, and many of them live permanently on land that is also used by rural communities for crops and grazing. As elephant populations grow, these two communities inevitably encounter each other, and sometimes those meetings erupt into violent conflicts.

Western images of elephants are informed by wildlife programmes, cartoons, films and books. We imagine that when we see an elephant it will be a glorious experience, a magical encounter while on safari in a wilderness area. When I first went on doctoral fieldwork to Zimbabwe I became rather irritated with friends and family, who kept asking 'Have you seen an elephant yet?'; my answer was always, 'No, I'm working in the capital city, Harare, and there are none here.' I was so focused on telling people that you had to go to national parks to see elephants that I was surprised when I finally did see an elephant, and it was far from the usual kind of tourist experience.

While on fieldwork, I had gone to Victoria Falls for a short break and I was lucky enough to be there during a full moon when the Parks Department ran its 'lunar viewings' of the Falls. I enjoyed the lunar tour, and to get back to my campsite I had to cross the Falls car park, cross the railway tracks and then walk up the main street. I got to the car park and saw some other tourists slowly driving off without any headlights on. I could not understand what they were doing. Then out of the darkness a park ranger came running towards me – in a loud stage whisper he asked me where I was going. When I replied that I was going back to my campsite, which involved crossing the railway lines, he said, 'Can't you see the elephant?' I looked, and I could not see anything – just darkness and the outline of trees against the night sky. Then I heard a branch snap and suddenly the tree seemed to come alive and thrash about, and in that moment I saw it – a huge male elephant. He was a regular visitor to the car park at night, keen to raid the bins for tasty leftovers from tourist picnics. I could easily have walked right into him, and would have been lucky to tell the tale. So my first sighting of an elephant was not the imagined magical wilderness experience, dressed in safari gear and brandishing binoculars or a telephoto lens, but in a car park, with people all around and the rank smell of bins. The incident served to underline to me that elephants live alongside people, and rural African communities have to deal with the challenges posed by elephants every day of their lives.

Southern Africa has the largest elephant populations in the world. Estimates vary because elephants are migratory and can be 'double counted' by more than one country; also some states have a political interest in inflating their elephant numbers to put pressure on CITES to allow trading or to justify culling. It cannot be doubted though that Southern African states do have the largest and most stable (even growing) populations in Africa. In 2008 South Africa claimed 18,000 elephants, Botswana estimated that it had a massive 190,000 elephants, Zimbabwe said it had 85,000 and Mozambique claimed 27,000. Marthinus van Schalkwyk, Minister of Environmental Affairs and Tourism for South Africa stated that the elephant population was increasing by 6–7 per cent per year, which was causing substantial management problems.[25]

It might be seen as an incontrovertibly good thing that elephant populations are growing in Africa, but it does produce problems. Whether we like it or not, African states cannot provide unlimited 'wilderness' for wildlife to roam; for better or worse people are actively involved in managing wildlife whether they live within national parks or roam outside them. In countries like Botswana and South Africa, if elephant populations are confined to national parks they strip out vegetation, which has negative impacts on other wildlife. It is easy to see where a large herd of elephants has been: a 'mini war zone' is how one former national parks director once described it to me. He was right: branches are broken, trees uprooted, vegetation trampled under-foot, waterholes fouled with elephant dung. In many ways this is a normal and important part of any ecosystem: the elephants create disruptions which then allow other plants and animals to prosper. The problem is that once the elephant population gets too large it destroys bigger and bigger areas, and then other wildlife gets forced out, including rhinos, which are much more endangered than elephants.

As elephant populations grow they increasingly stray outside national parks in search of food and water. Villages provide easy pickings, and elephants are partial to the crops grown by rural communities. Villagers will tell you about how they have had to fight off crop-raiding elephants, or how elephants have eaten all their crops. Villagers may spend months tending their crops to ensure a good harvest that will see them and their family through the year. Then, in the space of just a few minutes elephants can destroy a year's food supply. Rural communities all over Africa will happily show you the kinds of defences they have tried against elephants. These include draping washing lines with flut-tering flags which act as a giant scarecrow and guarding their crops all night, while hoping that they will not come face to face

4.3 Flags to scare elephants, Okavango Delta, Botswana (2008)

with a hungry elephant. Some have even burned packets of chilli powder because elephants dislike the irritating smoke; other tried and tested methods include using flaming torches or banging metal pots. It is hard for Westerners, sitting comfortably in their houses and accustomed to shopping for food rather than growing it, to understand that elephants can be so destructive and have such an impact on the ability of people to feed themselves.

But we must also remember that rural communities in Africa do not simply 'hate elephants'; this is an easy stereotype which is often used by those in favour of opening an ivory trade, but it also easily fits in with the ways some conservation NGOs present African communities as the enemies of wildlife. Most African societies have myths and legends about the origins of elephants, their importance to the ecosystem and their relationships to people.

For example, Lorraine Moore's study of rural communities in the Caprivi Strip in Namibia revealed that both children and adults loved elephants and would not like to see them disappear from the landscape.[26]

The 1989 ban coincided with the development of new experimental approaches to conservation that worked with local communities, so-called Community-Based Natural Resource Management (CBNRM). As discussed in chapter 3, its rationale was to reduce human–wildlife conflict by giving poor rural communities a financial incentive to conserve wildlife in their area. This would mean that wildlife populations could be sustained outside national parks and supply a long-term and workable solution to the problem of isolated populations confined to protected areas. They allowed communities to use wildlife in their area to generate income for households, and for local projects such as drilling wells or building schools. These schemes primarily raised money through the development of small-scale tourism, sport hunting and sales of wildlife products – and ivory was identified as one of the most valuable commodities. The idea was that in areas where elephant populations were healthy then communities could have a hunting quota and sell the ivory for their own benefit.[27]

Despite the clamour for an ivory ban Southern African states remained resistant, claiming that the trade could actually save the elephant, reduce poverty and produce revenue for conservation. For them, ivory sales could be a key component of their new CBNRM programmes, and would give people a financial incentive to conserve elephants. They underlined the fact that many elephants did not live in national parks, and rural communities had to bear the brunt of living with elephants. Southern African states also forcefully argued that their problem was that they had too many elephants, not too few; therefore they felt they had a right to determine how they should manage and use 'surplus' elephants; and they resented being told what to do by an international organization which included lots of member countries that did not have elephants and did not understand their management problems. This caused a rift between East and Southern Africa over how to approach elephant conservation, which still shapes the elephant debate and continues to produce bitter clashes over ivory.

1989: THE YEAR THAT SHOOK ELEPHANT CONSERVATION

By the end of the 1980s the globally dominant view was that elephants were in peril, and that they were being decimated by a senseless slaughter to feed the ivory trade. Spearheading the campaign against ivory was Dr Richard Leakey, then Director of the Kenya Wildlife Services. It was lost on no one that he was white, and his appointment chimed with a wider international grouping of white conservationists speaking 'for Africa'. What Leakey did have was an international profile, a network of supporters all over the world and the charisma and ability to mobilize them for his cause. Leakey was very smart at international networking and using the media. He appeared on US television and approached the World Bank and national governments of the US, Britain, Germany, Italy and Japan; he raised US$150 million for a five-year conservation programme and secured promises of a further US$150 million if it was successful.[28]

Leakey also masterminded an ingenious publicity stunt. In July 1989 Kenyan President Daniel Arap Moi set light to 12 tons of ivory. The resulting bonfire, the design of a pyrotechnist who specialized in creating dramatic fires for movies, represented the tusks of 2,000 elephants, and at the time it would have fetched US$3 million on the open market.[29] The photograph of Richard

Leakey standing in front of the burning ivory is an iconic conservation image. Leakey invited diplomats, conservationists, international media and government officials to watch the event, and also hired a public-relations firm from Washington called Black, Manafort, Stone and Kelly to ensure that it got maximum international attention. Leakey had calculated that the ivory bonfire, coupled with his crackdown on corruption in the KWS, would produce financial rewards from NGOs and donors. The gamble did pay off: donors and NGOs felt reassured that they had found someone who was really committed to saving elephants.[30] Kenya quickly became a favourite of the Western world, renowned for its commitment to conservation and its attempts to save elephants from poachers. The downside for local communities was that the outside world knowingly provided the resources needed to implement conservation policies that separated wildlife from human communities, and the controversial shoot-to-kill policies. But the Western love of elephants simply seemed more important; the images of burning ivory resonated with the public imagination in North America and Europe.

Kenya and Tanzania had felt the brunt of the poaching wars of the 1980s, and it is certainly the case that their elephant populations were decimated by illegal hunting to meet the international demand for ivory. This understandably meant that they were very much in favour of a global ban on the ivory trade to help them conserve their elephant populations in the longer term. But the view even from Kenyan conservationists was not as unified as Leakey and his supporters suggested. A former director of the Kenya Wildlife Services, Perez Olindo, remained in favour of a controlled ivory trade. He had been trained in the African Wildlife Leadership Foundation college, he was well respected in conservation circles, and he was a member of the IUCN African Elephant and Rhino Specialist Group.[31] Olindo saw Leakey's ivory bonfire as a colossal waste of an important resource. His view was that the elephants were already dead, and the tusks could have been sold – in fact ivory buyers had made offers for the stockpile. Ivory was an important source of income, and it was one which could (theoretically) be ploughed back into wildlife conservation.

But the publicity created by the ivory bonfire helped to galvanize international opinion in favour of banning the ivory trade. By the middle of 1989, world opinion had turned against the notion that a controlled and legal ivory trade could raise revenue for conservation, and demand for ivory in the West had all but disappeared. However, the heated debates about the rights and wrongs of ivory and the fears that elephants would become extinct meant that some of the complexities of poaching and trading were skipped over. The ivory bonfire was immediately met with criticism from Southern African states, which saw it as a crass publicity stunt which effectively silenced their alternative views. Such countries were adamant that ivory trading could work for elephant conservation in the longer term if it were controlled properly, and that proceeds from the sales of ivory stockpiles could be ploughed back into conservation and help to save elephants in the longer term. They argued that banning the ivory trade failed to deal with the real issues (demand, trafficking, organized crime) and merely punished the developing world for the failings of enforcement in the places where ivory was in demand, thereby making the poor even poorer.[32]

An important trigger factor in the move towards a ban was the early publication of a report by the Ivory Trade Review Group (ITRG), carried out by a consortium of NGOs but commissioned by CITES. Crucially, the ITRG report concluded that a ban was the most effective way of saving the African elephant; there was no real surprise there, but it was published while the ivory ban campaign was at its peak. Here it seemed was a group of 'experts' committed to saving elephants, and they were saying the only answer was a trade ban. But what is often forgotten is that the report also concluded that elephants in Southern Africa were in no way endangered by the trade.[33] Once the report was out in the public domain the US, EC, Switzerland, Australia, Canada and the UK on behalf of Hong Kong all announced full or partial ivory bans. Declaring a ban was a way of sending an international message that these countries were serious about saving elephants, and it played well to electorates at home.

The Ivory Ban

The crucial meeting was the Conference of the Parties in Geneva in October 1989. The debate at CITES revolved around three proposals: a total Appendix I listing (a ban), a split listing which would allow Southern Africa to continue with a controlled trade, or letting the ivory trade continue. Eugene Lapointe, the Head of CITES at the time, said that he felt the meeting was the stage where the animal-rights movement exploded into international consciousness. His view was that their arguments dominated the decision making at CITES and that ten or twelve NGOs turned the conference into a total farce, and he accused them of engaging in blackmail.[34] For example, delegates were greeted by a demonstration organized by WWF-Switzerland. The demonstration was unusual: no placards, no marching, just school-children wearing papier-mâché elephant masks, performing little sketches and reading poems they had written about elephants. The spectacle of impassioned Swiss schoolchildren pleading for elephants was a gift to the international media.[35]

Contrary to what we might assume, WWF was far from united over the ivory issue. WWF-International remained in favour of a controlled legal ivory trade right up until 1989, but WWF-UK and WWF-US broke ranks and started to call for a total ban on the trade. John Hanks, a prominent conservationist and head of the WWF-Africa Programme, was closely associated with the anti-ban stance of WWF-International. Hanks was in favour of sustainable use through a controlled ivory trade, and argued that the funds raised could be ploughed back into conservation in Africa. At one point during the run-up to the ivory ban WWF came to a compromise – that they would support what was called a 'split listing'. This meant that ivory trade from East Africa would be banned, but a controlled ivory trade would be permitted from Southern Africa, which had better managed and more robust elephant populations. Despite divisions in WWF over the ivory trade, the organization still got a lot of the credit for the decision to ban the trade, much to the irritation of other NGOs which had vocally campaigned for the ban.[36]

WWF, like many conservation organizations, is afraid of losing its fundraising base, and charismatic species like elephants, tigers and rhinos are big money-spinners for them. The public respond to appeals to save pandas and elephants more readily than they will respond to appeals to save insects or trees. We can understand why NGOs courted the public and played on emotions to raise funds, but in listening to its membership base, WWF ignored the opinions of the people and governments that had to live with and manage elephant populations in Africa. Some NGOs used shock tactics to convey their message and denounced those who opposed them as unconcerned with the plight of the elephant. This is not unusual among campaigning organizations which recognize that creating an air of urgency through shock tactics is the only way to get their particular issue into the public mind and onto the political agenda. They use exaggeration to play on our emotional responses to seeing orphaned baby elephants, or our revulsion at pictures of butchered elephants – which make a direct link between human greed, demand for fashionable trinkets and the violent deaths of one of the world's most iconic species. NGOs are aware that the public respond to these images and slogans, and that it is much harder to sell a complicated story about variations in ivory production, the organization of poaching or the importance of engaging local communities.

The NGO campaigns created an atmosphere where it was almost impossible to challenge the idea that a ban was the best way forward – to do so was suicide for any conservation NGO. The CITES vote was seventy-six in favour of a ban, eleven against and four abstentions. Formal ivory trading ceased on 1 January 1990, and shortly after this, the carving centres in Hong Kong and China virtually shut down.[37] The speed of the collapse in the ivory trade caught many conservationists by surprise. NGOs were able to claim a real moral victory – they had saved the African elephant from extinction. The legacy of the ban for the WWF was more complicated; the organization actually split over it, so WWF-US and WWF-UK are now separate from WWF-International.

The Legacy of the Ivory Ban

The decision to ban the ivory trade in 1989 was not the end of the story. Instead, it was the start of bitter wrangling aimed at satisfying divergent views on how to save elephants. In the interests of keeping the members of CITES together and managing a common position on ivory it was clear that some kind of compromise was necessary to prevent Southern African states simply leaving CITES altogether. The compromise was that countries could take out a 'reservation' on the ban. This allowed them to apply to downgrade their elephants to Appendix II if they could prove they were managing elephants well. This at least gave Southern African states some hope that in the future they would be able to reopen a legal ivory trade if they could prove they had managed their elephants well. Zimbabwe, South Africa, Botswana and Namibia continued to hold reservations and were allowed to trade ivory internally, within wider Southern Africa and with any state that was not a member of CITES. This lead to some curious scenes in tourist trinket shops in the Southern African region. On more than one occasion I stood in souvenir shops in Zimbabwe in the mid-1990s to hear tourists from all over the world gasp in horror at the ivory goods on sale. The tourists assumed that the shop was selling ivory illegally when in fact it was perfectly legal for citizens from any Southern African country to buy ivory and take it home with them – it was only illegal for the rest of us.

4.4 Ivory on sale illegally in New York (2008)

Rowan Martin, the deputy director of National Parks in Zimbabwe, complained that conservation NGOs had been too influential. His view was that they had blackmailed donors into devising aid packages with 'elephant strings' attached. Here again was an example of elephant-free Western states telling African states how best to manage their elephants, which for some Southern African conservationists smacked of colonialism. Rowan Martin commented that at the 1989 meetings he was approached by the US delegation and was reminded that he should think of the aid that Zimbabwe received from the US and World Bank when voting on the ivory issue.[38] Indeed, one conservationist I interviewed in Zimbabwe in the mid-1990s commented that the American delegation at CITES could be 'really mean' if they did not get what they wanted. The feeling that the international community misunderstood the Southern African position was particularly acute because the region had a relatively good record on poaching in the 1980s, and it was felt that Southern Africa was being punished for East African elephant management failures. This sentiment translated into charges of neo-colonial behaviour against those in favour of an ivory ban.

As a result of the decisions at CITES in 1989, national parks and wildlife authorities in Southern Africa have to deal with the growing elephant populations in ways that are in line with inter-national law. These days Zimbabwe is better known as a country in crisis, destroyed by dictatorial rule which has produced wide-spread human suffering including cholera epidemics, poverty and malnutrition. But in 1989 the story was a little different. At the time Zimbabwe could boast of one of the most professional parks departments anywhere in Africa, with a well-run system of national parks and a healthy elephant population. The parks staff were internationally renowned conservationists, including Rowan Martin and Brian Child. They knew they were definitely part of what is called the 'khaki shorts brigade' – a reference to the white community in Southern Africa's characteristic costume of khaki shorts and shirts, knee-high socks and boots. But they were acutely aware of the need to engage local communities and in the 1980s they became unlikely champions of black empowerment. They were very much in favour of engaging local communities in wildlife management, and in allowing them to manage their wildlife populations sustainably. They argued that this was the only way to ensure a long-term future for wildlife, and especially elephants, because parks could not provide enough space. In addition they saw that international support for elephant conservation was being concentrated in East Africa, so they needed to find a way to pay for conservation that did not rely on international donors and NGOs. For them the fate of elephants lay in the hands of the very communities that many conservation NGOs blamed for the elephants' decline.[39] The Parks Department view was that the people should be allowed to manage wildlife in their area, and be offered financial incentives to do so.

As one Zimbabwean Parks Department official put it to me in the mid-1990s, if communities decide to kill all the wildlife in their area they have the right to do that and we will have to accept it because they live with it, but he stressed that such an outcome was highly unlikely. His view was that communities were not stupid and they could see they would benefit from sustainably managing their wildlife.[40] Sometimes this is referred to as an approach which turns 'poachers into game guards', but in Zimbabwe they went one step further – local communities did not just act as protectors of wildlife, they became wildlife managers.

In the case of elephants this meant they argued in favour of continuing with the ivory trade. The Parks Department even devised a model conservation programme which was copied, modified and improved across Africa and further afield. The

Communal Areas Management Programme for Indigenous Resources (CAMPFIRE) allowed rural communities to determine how they might profit from using wildlife; this included selling permits to allow trophy hunting on their land, distribution of meat from culls and the development of photographic safaris. The first area to be declared an 'Appropriate Authority' to run the scheme was Nyaminyami District in 1989. The decision followed years of developing the idea of CAMPFIRE and running smaller-scale schemes such as distributing meat from impala culls carried out by the Parks Department.[41] The programme was quite successful in some areas, generating cash and other wildlife products to rural communities. This programme was the centrepiece of Zimbabwe's claim that CITES and those in favour of a ban might think they were saving elephants, but their decisions were negatively affecting some of the worlds' poorest rural communities. At the end of the day the decision to ban the ivory trade meant that rural communities in Zimbabwe could not sell the ivory produced in their area; they could only derive benefits from allowing elephants to be sport hunted under a government quota, or through the development of photographic tourism which is unsuited to many of the areas because of low densities of wildlife, difficult terrain or lack of tourism infrastructure. It seemed clear to them that the decision to ban the ivory trade may have brought short-term benefits to elephants in East Africa, but meant that Southern Africa would be further impoverished, reducing local support for conservation.

CULLING AND THE STOCKPILES CONTROVERSY

Following the ban in 1989, ivory began to pile up in Southern Africa. It came from confiscations of illegally hunted ivory, from culling operations, from natural deaths of elephants and also from 'Problem Animal Control' (PAC). The growth of ivory stockpiles has been used by Southern Africa to argue that a potentially valuable and renewable natural resource is simply being wasted as a result of an ill-informed international decision at CITES. Right from the start of the ban Zimbabwe had to manage a growing stockpile of ivory. When I met Rowan Martin in 1995 at the Parks Department HQ, he claimed that he was being approached by Asian dealers interested in the ivory held in the central store, and that generally offers were around US$1000 per kilo, which meant that the 30-ton stockpile was worth US$30 million at the time.[42] He argued that the ivory could bring in much-needed funds for conservation and saw it as an unacceptable waste of a renewable resource. The ivory stockpiles in Southern Africa came from a range of different sources. Some of the more obvious sources were confiscations of illegally poached ivory, or ivory from elephants that had died of natural causes. However, there are two other more significant sources of ivory which continue to divide conservationists.

The first controversial source of ivory in national stockpiles comes from culling operations. Culling has to be one of the most divisive issues in conservation and it is an issue that fires up elephant specialists almost as much as the ivory trade does. Since we tend to assume that the elephant is endangered across Africa it is hard to accept that some countries might consider deliberately reducing their elephant populations. But Southern African states have all considered culling, and some have even carried out culls in national parks when the elephant populations have reached a level where they are damaging the landscape and other wildlife. Southern African states fear reprisals from the international community if they announce a cull; understandably, they worry about the impact on their profile in conservation and how it might damage their tourism industries.

Animal welfare-oriented NGOs such as the UK-based IFAW and the Humane Society of the US (HSUS) have vehemently opposed culling as cruel and completely unnecessary. They claim that animals are distressed by the cull, which involves shooting entire family groups from the air. The South African Parks Department used to cull specific individuals within a herd but found that this disrupted the structure of the herd and caused problems in the longer term; their position is that while it may be hard to accept the sight of baby elephants being shot in culling operations it is actually more humane to kill entire family groups to minimize disruption.

Botswana and South Africa have temporarily suspended culls because of fears of reprisals. But in 2008 the Minister of Environmental Affairs and Tourism Marthinus van Schalkwyk announced that South Africa would resume culling in Kruger National Park for the first time since 1994. The elephant population in Kruger had swelled from 8,000 to more than 12,000 and the numbers of elephants were destroying the landscape and putting other species at risk.[43] IFAW and HSUS have insisted that the South African Parks Department consider the alternatives to culling, and deny that culling is ever justified. Critics of these organizations point out that while they proclaim the rights of elephants they have failed to offer sufficient funds to support alternatives to culling. For Southern African states this is further evidence of Western hypocrisy and double standards: these organizations expect some of the world's poorest states to pay the costs of keeping large elephant populations, while they are prevented from gaining the revenue from ivory sales. Meanwhile, the international community offers little financial support for wildlife management.

The second controversial source of ivory held in Southern Africa's stockpiles comes from 'Problem Animal Control'. PAC operations deal with elephants that have been identified as 'rogues'; these are elephants that persistently raid crops or threaten human life. In PAC operations, rural communities identify an animal that has become a pest so the parks department can shoot it; alternatively a licence to kill the PAC elephant can be sold to a sport hunting safari operator, and then international sport hunters will pay to shoot the animal. Sport hunting is a very controversial activity, but it can also be a very lucrative business. For example in 2009 in Zimbabwe a seven-day package to sport hunt a PAC elephant was being sold for US$6,900 (hunting of ordinary non-PAC elephants is more expensive, at around US$20,000). PHASA estimated that in 2008 sport hunting of wildlife earned South Africa a total of US$100 million per annum.[44]

For many conservationists the idea of sport hunting is abhorrent. Many of us feel a sense of revulsion at the notion that someone might pay such vast sums of money to hunt such a beautiful animal. But the fact remains that properly controlled sport hunting is a sustainable and highly lucrative industry which generates millions of dollars a year for some of the world's poorest states. Sport hunters draw attention to the inherent hypocrisy in the arguments made by animal welfare NGOs and armchair conservationists in the West; they champion the rights of individual elephants but happily ignore the harsh realities of human poverty and of overburdened national budgets. Opponents of PAC and sport hunting counter this by arguing that the extra funds generated are not necessarily ploughed back into conservation and rural communities; instead, corrupt governments and wealthy safari operators are the real beneficiaries.

While conservationists remain divided about whether PAC, sport hunting and culling are useful for conservation, there have been attempts to find alternatives. South Africa has explored the possibility of 'translocating' surplus elephants to other countries that have fewer elephants, such as Mozambique. There are professional companies that specialize in capturing and moving wildlife, but so far the high costs have meant that it has not provided a long-term solution. Southern African states have called for international conservation NGOs to fund translocations of surplus elephants, so that culling is no longer necessary. But so far parks departments have been left to pay up on their own.

Another potential solution was tried in Kruger National Park, where female elephants were darted with a contraceptive to reduce the population. However, this also proved to be very expensive; there were also logistical problems when parks staff discovered that in order to maintain the effectiveness of the contraceptive they needed to locate and dart the same animal one year later. The contraceptive route also raised an unforeseen animal welfare issue. Using contraception means female elephants come into heat every four months instead of every four years; this exposes them to considerable stress from multiple matings per year.[45] Therefore, although some alternatives have been tried, they have either raised new problems or simply been too costly for Southern African governments.

Debates about translocation, contraception, culling, PAC and sport hunting are always related back to the core problem, the international ivory ban. The ivory stockpiles have proved to be a really thorny issue for CITES. For Southern African states these stockpiles seem like a colossal waste. They draw attention to the hypocrisy of the international community, which is clearly against culling but unwilling to pay the immense costs of translocation or compensation to destroy the ivory stockpiles. The response of organizations like IFAW is that corrupt African governments will not funnel the funds back into conservation, but will pocket the money for themselves. This chimes with stereotypes of African people and governments who are presented as part of the problem rather than part of the solution. Anti-ivory trade groups also argue (perhaps more convincingly) that a legal ivory trade will only act as a cloak for a more damaging and larger scale illegal ivory trade, and the risk to elephants is far too great.

CITES has really struggled with the issue of stockpiles and growing elephant populations in Southern Africa; in response the complete trade ban has been relaxed since 1989 to allow one-off sales of ivory stockpiles. The first sale was in 2004, and the most recent was October to November 2008 to approved Chinese and Japanese dealers. Namibia sold 9 tons and raised US$1.2 million, while Zimbabwe made US$480,000 and Botswana received US$1.1 million.[46] Richard Leakey publicly denounced the sales and criticized CITES for allowing China to join the auction because the country scored so badly on the CITES mechanism for monitoring the ivory trade (the Elephant Trade Information System, or ETIS).

Leakey's press statement did not contest the claims that Southern Africa had good elephant management plans (with the exception of Zimbabwe) but he did reiterate the need for Southern Africa to think of elephant populations on the rest of the continent. He stated: 'I believe that auctioning the ivory stockpiles would cause poaching to increase particularly in the central, eastern and western African elephant range states where poaching is not yet properly controlled.'[47] The slight relaxations of the ivory ban since 1989 still cause serious divisions, but so far have been successful at keeping Southern African states within the convention. This is very significant; it would critically weaken CITES if the states with the largest elephant populations in the world simply left the convention and began to trade outside it. The issue of reopening the ivory trade is far from over; in fact it looks set to remain on the conservation agenda for some time to come, and the legal ivory sales in 2004 and 2008 permitted by CITES are important test cases. As it stands, the global ivory trade is much smaller than it was in the 1980s, but some conservationists worry about a return to the destructive poaching wars of the 1980s, which filled our TV screens with distressing images of butchered elephants.

The core of the bitter disagreements over ivory can be found in two different approaches to wildlife management: preservation versus sustainable use. At its most basic, preservation revolves around saving or preserving animals, usually in national parks. In the strictest sense preservationist approaches mean no use of wildlife at all, not even for tourism. This model is reflected in the definitions that IUCN use for protected areas; for example an IUCN Category Ia area is a strict preserve which can be used only for scientific research, while a category Ib is a wilderness area managed to protect its 'natural condition'.[48] In contrast, sustainable use does allow for human use of wildlife as long as it does not damage long-term survival. The problem is that 'use' is a rather fuzzy term. Conservationists have tried to grapple with this by defining different uses of wildlife as either 'consumptive' or 'non–consumptive'. This may sound like some sort of Orwellian doublespeak, but the terms cause deep divisions in conservation. Non-consumptive uses of wildlife are basically those that do not kill the animal. The most common example of non-consumptive use is the development of photographic tourism. Consumptive use refers to activities which result in the animal being killed. Sport hunting is a good example of this: a small number of animals are identified in a quota system to be shot by commercial safari hunters.[49] A controlled ivory trade is another potential example because the elephant has to be killed to extract its ivory. The idea is that a small number of animals will be killed to generate finance for conservation of the wider species. These differences are critical and underlie the deep divisions over whether a controlled ivory trade is responsible for wiping out elephants or whether the trade is a potential solution for elephant conservation.

In the debate on the advantages and disadvantages of a legal ivory trade, the problems with enforcing the ivory trade ban and the complex ethical issues raised by the relationships between elephants and rural communities brings the main arguments of this book into sharp focus. The ivory debate reveals the hard 'real world' problems of stamping out wildlife smuggling, but it also shows us how conservation rules define, create and sustain various forms of criminality. More controversially, the harshest critics of the ivory ban claim (with some justification) that it simply makes the poor poorer.

The splits over whether it is ethical and permissible to kill elephants to trade their ivory is informed by vastly differing views on elephants. Elephants are an important 'flagship' species for conservation NGOs, who use them as fundraisers. Their appeal in the West comes from the ways they are regarded as sensitive, emotional, intelligent and family-oriented animals; they are the subject of cartoons, children's books and films, and images of elephants can be found on almost anything from mugs to birthday cards to t-shirts. To Western eyes, elephants are special, almost human. These qualities endear them to the legions of armchair conservationists in the worlds' wealthiest states. Parks departments, especially in Southern Africa, fear NGO campaigns that denounce their position on ivory because it could damage their international reputation or, more importantly, their lucrative tourism industries. Elephants remain at the centre of arguments over the hypocrisy of Western conservationists: that they care more about animals than people and that they do not understand the harsh realities of managing large wildlife populations.

4.5 Ivory on sale illegally in Ethiopia (2002)

Those in favour of the ivory ban argue that community-operated schemes can still work perfectly well without revenue from ivory. Tourism has been promoted as an alternative pathway for using elephants sustainably and humanely. Safari tourism is certainly a big earner for some parts of Africa, especially Kenya and Tanzania. But it is important to remember that in most rural areas of Africa, tourism infrastructure is very underdeveloped and some countries are not on the international tourism map. Wildlife populations in many areas are also not dense enough to support international safari tourism. Tourists want guaranteed sightings of herds of elephants, and they can easily get this in world-famous national parks like Amboseli, Ngorongoro and Kruger; they would be less likely to pay the same for a safari experience which could not promise wildlife and which offered only basic transport and accommodation facilities. As a result, tourism is simply not the answer for many rural communities, and is in fact a red herring which distracts from the harsh realities of long-term elephant management. Ultimately, critics of the ban see it as a temporary band-aid which does not offer a long-term solution to the main problem: how to get people and wildlife to live together. As a result, it is regarded as a highly unjust imposition on the poorest communities by some of the richest communities in the world.

CONCLUSION

The continuing wrangles over ivory and rhino-horn trading reveal that simply banning the trade does not provide a simple and effective answer. Illegal trafficking continues, and controlling the legal ivory trade remains a challenge for law enforcement agencies. Trade bans are problematic because they are a one-size-fits-all solution – a simplistic and blunt instrument which fails to grapple with the everyday complexities of wildlife management in different places all over the world. The Western love of elephants is shaped by sentimental ideas about them as human-like, but fails to appreciate the less romantic everyday realities of living with elephants. The best example is the failure to understand why rural communities might be hostile towards elephants that raid crops, or why they may want to be compensated for loss of revenue from ivory trading. If conservationists do not tackle the ways elephant conservation intersects with rural poverty, human–wildlife conflict and the constraints on national budgets then the elephant will be doomed in the longer term. Hoping that safari tourism will provide the answer is hopelessly naïve because many elephants live in areas unsuited to development of international tourism.

Southern African states argue that a controlled ivory trade might just save elephants in the longer term. It can provide revenue for conservation and rural development, allow local communities to realize the full value of elephants and give them incentives to manage wildlife in their area. This too might seem naïve, especially since the history of relationships between rural communities and states in Africa is chequered, and is replete with examples of how governments ignore the needs of rural areas. Either way there is a clear injustice being perpetrated here. Some of the world's poorest communities are prevented from using wildlife in a sustainable way for their own benefit. This might be acceptable if the international community actually provided the resources to implement alternatives, such as buying up stock-piles at a reasonable price or funding elephant translocation schemes. But no such support has been forthcoming. Instead international decision making, informed and directed by Western interests, ensures that local needs, desires and viewpoints are absolutely invisible. This will inevitably undermine local support for conservation, which is needed to secure the elephants' fate in the longer term. Wildlife is saved so that wealthy Westerners can visit them on holiday or watch them on TV. Trade bans are hailed as conservation successes, but the really negative and unjust consequences are glossed over.

CHAPTER FIVE

GUERRILLAS TO GORILLAS: BLOOD DIAMONDS AND COLTAN

As I SAT IN the Travellers Rest hotel near Kisoro in Uganda, an American tourist said to me, 'Seeing the gorillas is amazing – it will change who you are *forever*.' I was mentally preparing myself for a challenging hike the next day to see the rare mountain gorillas in the Nkuringo section of the Bwindi Impenetrable National Park. This was not the first time tourists had said this to me. Perhaps there is something special about gorillas, something that touches us all, something that changes us. But undertaking a trip to see the mountain gorillas is not for the faint-hearted. Everyone and everything from tour guides to guide books and magazines repeatedly warned that it could take hours to find a family of gorillas, and that at the end of an exhausting day of gorilla tracking you might well face a four- or five-hour trek up the steep slopes to get out of the valley. Add to that blazing heat, torrential rain, the giant elephant nettles and paths so slippery they can seem like ice. After seven hours of wading through thick rainforest, sliding down steep valleys, and scrambling up mountains, I was exhausted, but I had spent one hour watching a gorilla family playing, eating, sleeping and fighting. Was I changed by the experience? No, but I enjoyed it, so an unchanged me did it all over again two days later in Rwanda.

Despite the obstacles, the parks services in Uganda and in Rwanda have no trouble finding willing buyers for their gorilla tracking permits, which currently cost US$500 per person per visit. The question is why gorilla tracking is a major tourist attraction especially for Rwanda and Uganda. It would be too easy to assume that their popularity stemmed from work by people like Dian Fossey, or the famous images of Sir David Attenborough interacting with gorillas in Rwanda. Perhaps, like elephants, there is something special about mountain gorillas that touches us. Some of my fellow gorilla tourists thought it was because they were closely related to us, and that when you looked in their eyes you felt a connection with them. This view of gorillas is a world away from the earlier vision of them as monstrous beasts, as a violent 'King Kong'.

Gorillas are big business; they generate a lot of money for governments, tour operators, lodge owners and a whole raft of souvenir and craft sellers. In Rwanda and Uganda they are the jewel in the crown for the national parks services, which use the lucrative permit sales to support conservation in other protected areas.[1] Mountain gorillas are also found in the Democratic Republic of Congo (DRC), but a recent history of conflict there means the gorilla trekking industry remains small. Mountain gorillas are big business for conservation NGOs as well; like elephants, they are a flagship species and vitally important for fundraising.

The cry from conservation NGOs is that conflict is threatening both mountain and lowland gorillas: that illicit mineral and gemstone mining to pay for vicious wars in Africa is pushing these gentle giants to the brink of extinction. But as in other

5.1 Welcoming tourists to the Parc National des Volcans, Rwanda (2007)

areas where wildlife conservation and criminal activity meet, these claims are oversimplifications. There is an indisputable link between conflict, mining and wildlife conservation, but some of the responses to this have made matters worse for wildlife and the people who share their habitats. Mainstream conservation thinking tends to focus on defending wildlife on the 'front line' from illegal miners and poachers. But this is part of the problem, because it fails to confront what drives poaching and illegal mining, and that is demand in the rich world.

As a result, conservation strategies are devised which tackle the symptoms rather than the underlying causes. Demand for diamonds, sapphires and strategic metals for the new mobile technologies that so many of us rely on is rarely mentioned, let alone confronted. This means that the strategies designed to conserve wildlife affected by illegal mining and conflict fail to address the real challenges of global demand, and point the finger of blame in the wrong direction. The miners and poachers are simply presented as the enemy, leaving the consumers of electronics, mobile phones, diamond rings and sapphire earrings to continue with business as usual. This chapter highlights why it is important for conservation to respond to and understand international dynamics. It focuses on the global networks that drive and sustain conflicts and mining that threaten some of the worlds' most endangered wildlife. Here I delve into the simplified picture of greedy poachers, poverty-stricken miners and violent warlords to unpick the heart of the problem: consumer demand for jewellery and mobile technologies in the rich world.

THE GORILLA MURDERS

Since the mid-1990s, the DRC has been better known as a site of acute conflict. But DRC is also a vitally important wildlife habitat for some of the world's most endangered species. It has three important and endangered species of gorillas (western lowland, eastern lowland and mountain), which are all threatened in different ways by conflict. Their precarious position is made worse by ongoing conflicts in the central African region. Unless we understand these conflicts more fully we cannot begin to save gorillas in the wild. The conservation movement sees conflict as 'the problem' but fails to really understand it, making the situation worse for gorillas and the people of central Africa. The DRC is home to the only population of the eastern lowland gorilla (*Gorilla beringei graueri*), and it is thought that there are less than 5,000 individuals left. DRC shares the rare mountain gorilla (*Gorilla beringei beringei*) with Rwanda and Uganda, although studies show that this population has actually increased despite the ongoing tensions and conflict in the borderlands between the three states.[2]

In July 2007, the international press was awash with stories about the 'murder' of four mountain gorillas in the Virunga

5.2 Mountain gorilla, Parc National des Volcans, Rwanda (2007)

National Park in DRC; and a year later the story was still important enough to be the cover story of the July 2008 issue of
National Geographic and the subject of a programme, 'Gorilla
Murders', on the National Geographic television channel. The
bodies were intact, so they were clearly not killed for bushmeat,
nor were they killed so their heads and hands could be sold as
trophies; it was unlikely they had been killed to take their babies
for the international pet trade since the baby was left unharmed
and found clinging to its dead mother's body; and it was also clear
that they were not killed in an attempt to protect crops because
they were nowhere near a human settlement. The crime
appeared motiveless and wildlife NGOs were at a loss to explain
why the animals had been killed. The deaths even prompted an
investigation by the UN agency UNESCO. The case raised
vitally important questions about what drives poaching and who
exactly is involved in it. While the initial reports claimed that the
attack came out of nowhere and was inexplicable, irrational and
motiveless, later reports revealed a much more complex situation
which encompassed warlords, local communities, rebel militias,
charcoal smuggling and illegal cattle grazing.

Conservation NGOs and the international media portrayed
the gorilla killings as unexpected and surprising, an irrational
and spontaneous act. The killings took place in a complex cross-
border area where the boundaries of Uganda, Rwanda and DRC
meet. While the gorilla 'murders' gained international press
attention, the human casualties of the ongoing violence in DRC
rarely get a mention. International conservation organizations
working in the area refer to a pointless war which threatens gorilla
habitats, thereby silencing and ignoring the ways international
demands for coltan (i.e. the ore columbite-tantalite), diamonds
and timber sustains and fuels the conflict. The Frankfurt
Zoological Society and Wildlife Direct (run by Dr Richard

Leakey) work in the area alongside government agencies such as the Congolese Institute for Nature Conservation (ICCN).

Wildlife Direct persistently complained that Laurent Nkunda, leader of a militia group in the area, had been moving cattle into the Virunga National Park, bringing his troops into conflict with government officials and with wildlife rangers. Nkunda also controls the areas used for coltan mining, which was listed in a UN investigation as a key mineral used to fund and sustain conflict in the DRC.[3] Coltan attracts a high price on the international market and is used in mobile phones, game stations and numerous other electronic devices. Some estimates suggest that 80 per cent of the world's supply of coltan comes from these mines in the area under Nkunda's control, particularly the Kahuzi-Biega National Park. The park is also a critical eastern lowland gorilla habitat, more than 50 per cent of which population WWF estimated disappeared during the war.[4] Coltan mining is a real threat to gorilla habitat, and its production is directly related to global demand for electronics and the growing nanotechnology industry.

Wildlife in the central African region is used as part of a wider conflict strategy. Wildlife Direct had warned that militias operating in the area had issued a threat to kill mountain gorillas two months earlier. In early 2008, the rangers supported by Wildlife Direct speculated that the killings were part of a longer-term and deliberate strategy to send a message to conservation groups working in the area that their presence and harassment of militia members was not welcome. Park rangers had been destroying kilns and confiscating illegal charcoal because the trade was blamed for destroying gorilla habitats.[5] The charcoal trade, which meets the massive demand for fuel from towns and refugee camps in the Kivu area, is worth millions of dollars each year, far outstripping any potential revenue from future plans for gorilla-based ecotourism. Conservation organizations have placed the blame on local people for using charcoal, without demonstrating any understanding of how charcoal demand is related to conflict in the region; the conflict has driven people out of their villages and farms and into town and refugee camps where they are forced to be dependent on markets in charcoal and foodstuffs rather than being able to grow their own food or collect their own woodfuel. It was alleged that all three armed groups present in the area were involved in the illegal charcoal trade: Nkunda's militia, the remnants of the Interahamwe (in hiding since carrying out the Rwandan genocide in 1994) and the Congolese army. There have been attempts to develop alternatives to reduce pressure on gorilla habitat, and to cut rebel groups out of the commercial loop. In 2009 rangers in the Virunga National Park tried to persuade people to switch from charcoal to briquettes made from sawdust, rice husks, leaf mulch and other organic waste and installed 550 presses in the area to make the briquettes. Meanwhile, Mercy Corps, an NGO working with Internally Displaced People (IDPs) in eastern DRC, constructed 20,000 fuel-efficient stoves to reduce demand for charcoal.[6]

Therefore the gorilla killings, rather than being an irrational or spontaneous act of cruelty against 'gentle giants', were part of a longer-standing conflict over control of the parks and its resources by various warring parties.[7] Other commentators, more critical of Wildlife Direct, have suggested the killings were also part of a strategy of resistance of local communities against the arrival of highly trained and well-equipped rangers behaving as a military force. This aspect of the story was not covered by the international press, including *National Geographic* and the *Washington Post*; instead they seemed to accept Wildlife Direct's version of events, which was thus promoted internationally. Perhaps it was not a coincidence that the *Washington Post* and *National Geographic* both have board members who are also on the boards of international conservation organizations

(Conservation International and the Dian Fossey Gorilla Fund-International). While the international press picked up the story of the 'massacre' of four mountain gorillas they hardly mentioned the millions of Congolese people killed in over a decade of conflict. The gorilla killings did not appear out of nowhere but were in fact the result of the existing dynamics of rebel activity, ongoing conflict and divisions between conservation agencies, local communities, government troops and rebel movements over the rights to control land and resources inside the park.

Gorillas are one of the hottest conservation issues. Conservationists were quick to raise the alarm about the impact of wars on some of the world's rarest and most charismatic animals. Wildlife Conservation Society points to the serious challenges faced by conservation schemes in the region:

> While most conservation programs were suspended in the country from 1995–2003, WCS actually expanded its activities during this period of civil strife. Throughout the armed conflict, WCS-DR Congo staff often took the lead in calling attention to poaching by military factions, militia involvement in illegal mining, and the rampant exploitation of local populations. After the country held its first democratic elections in 2006, the government began to re-establish order and rebuild roads, and today strives to bring an end to chronic civil unrest in the east.[8]

This presents an oversimplified picture of the problem; it talks about conflict and genocide as though they appeared from nowhere. There is sympathy for local populations who are exploited by various militias, but the military are cast as poachers and militias as criminal miners – as though they are fundamentally bad and need to be stopped by an army of well-meaning conservationists. The messy causes of the conflicts are conveniently stripped out and made invisible to our eyes. This means the world is presented with a stark and stereotyped picture, with locals portrayed as black/bad/poachers/rebels, and conservationists as white/good/committed/saviours. Across the world, wildlife has been caught up in conflict and in its illicit trading networks that sustain warfare. But the linkages are more complicated than they first appear. Illegal mining to meet demand for diamonds and coltan in the wealthy world underpins and sustains conflict in the central African region, but this is rarely explained in full. When illegal mining is mentioned it tends to be presented as the result of the greed and corruption of rebel groups, national armies, government bureaucrats and local villagers, all of them black and African. Just as we saw with poaching debates, this presents a false and racialized picture of Africans as the problem; the Western world appears as a saviour, as conservationists intent on saving gorillas or as UN peacekeepers trying desperately to end a senseless and brutal conflict. This image strips the problem of its context. Here I will explain the driving forces behind the production of 'blood diamonds' and coltan which sustains the conflicts that threaten gorillas. This places the problems of poaching, mining, warfare and encroachment into national parks in their global context.

CONFLICT IN CENTRAL AFRICA: THE 'NOT SO NEW' WARS

Conflict has marked the central African region for over fifteen years. The DRC war had two phases between 1996 and 2003, and while the war officially ended in 2003, fighting has continued. Oxfam estimates that 5.4 million people were killed by conflict in the DRC between 1998 and 2008. However, this is likely to be a significant underestimate because of the difficulties of identifying war-related casualties and because so many deaths go unrecorded. In fact, Oxfam claims that the DRC conflict is the deadliest since the Second World War – but it hardly ever hits the headlines.[9] The current rebel activity and ongoing unrest in eastern DRC has its roots in longer-term regional dynamics. The Rwandan genocide of 1994 is often identified as the trigger for a much bigger regional conflict that was described as the world's biggest war in the mid-1990s. The trigger for the genocide was the death of the then Rwandan president Juvenal Habyarimana, whose plane was shot down above Kigali airport on 6 April 1994. Journalistic accounts of the genocide have tended to view it through a convenient 'ethnic lens' – that the genocide was caused by age-old tribal hatreds between Tutsis and Hutus. It is certainly the case that most of the people killed were ethnically Tutsi, although moderate Hutus were also targeted. The idea that the genocide was a spontaneous violent eruption caused by tribalism made it easier for the international community to turn a blind eye to another messy and intractable African conflict.

But the genocide was preceded by a series of events that produced a domino effect that ultimately led to a conflict that sucked in DRC, Uganda, Rwanda, Central African Republic, Angola and Zimbabwe, among others. In 1986 Yoweri Museveni's National Resistance Movement toppled Milton Obote's government in Uganda. Museveni was regarded as a part of a new generation of reformist African leaders, and he went against the established rules of the Organization of African Unity (OAU) of non-interference in the affairs of neighbouring states. Museveni was not averse to intervening in other states if they presented a threat to Uganda or to their own citizens. By 1990 this chimed well with the international climate, characterized by a post-Cold

War faith in the idea of military intervention to protect human rights or to engage in humanitarian relief.[10]

Museveni provided important support for Rwandans living in Uganda, who had fled from the country once a Hutu-dominated government came to power at Independence. They became the Rwandan Patriotic Front (RPF), which invaded Rwanda from Uganda in 1990 to gain leverage for negotiating a power-sharing agreement. This alarmed the Hutu-dominated government, and allowed a Hutu power movement to argue that unless the Tutsi threat was crushed they would take over and persecute Hutu people. This was a powerful argument because it tapped into fears that stemmed from the Belgian colonial period, when Belgian colonists decided that Tutsi were a more aristocratic people, and placed them in positions of power over Hutus in a classic colonial strategy of divide and rule. Hutu extremists played on fears that an RPF victory would see a return to the bad old days of colonial rule. With the active support of Agathe Habyarimana, the President's wife, the Hutu power movement was able to organize, plan and mobilize for the genocide. Her political clique, often referred to as the Akazu, was seen as an important source of power in the run-up to the genocide. As the RPF gained control of more and more territory, the United Nations oversaw peace talks in 1993 in Arusha, Tanzania, to develop a power-sharing agreement. The difficulty was that neither side was especially interested in power sharing: the RPF were winning control over Rwanda and the Hutu power movement wanted to gain control of the existing government.

President Habyarimana signed the power-sharing agreement in Arusha, but his returning plane was shot down as it approached Kigali airport. This was the signal to begin the genocide. The Hutu militia known as the Interahamwe were unleashed and began a programme of organized killings of Tutsi and moderate Hutus.

The genocide was definitely not a spontaneous outburst of violence driven only by some sort of deep-seated ethnic hatred. It was about power and control over the Rwandan state, and ethnicity was only one part of the story. Over a period of 100 days 800,000 people were killed in a highly organized genocide while the international community failed to agree on what to do about it.[11]

As the RPF gained control over more and more territory in Rwanda Interahamwe and thousands of Hutu refugees crossed the border into DRC (then called Zaire). This sparked a conflict in Zaire which toppled one of the longest serving dictators in Africa, President Mobutu Sese Seko, in 1997. Laurent Kabila and his Alliance of Democratic Forces for the Liberation of Congo (ADFL) seized power in 1997 amid backing from the international community, and especially from Rwanda and Uganda. This cosy relationship between Kabila, Museveni and Paul Kagame, the RPF President of Rwanda, soon soured, and the DRC was shattered into a number of different territories, where multiple rebel movements and warlords fought for control. Uganda and Rwanda each backed their own rebel movements which allied in eastern DRC, but these soon became difficult to control and began to fight each other. Serious tensions even arose between former allies Uganda and Rwanda, and they threatened to go to war with one another over clashes and mining interests in DRC. Later in 2001 Laurent Kabila was assassinated by one of his own bodyguards, and was succeeded by his son, Joseph, in Africa's first example of a royal-style succession. Joseph Kabila was more interested in negotiating a withdrawal of foreign troops from eastern DRC. Apart from proxy rebel movements, Rwanda also had troops stationed within eastern DRC; Kagame claimed that they were there to prevent a reinvasion of their country by the remnants of Interahamwe who had disappeared into the refugee camps and rainforests of the region.[12] The troops were withdrawn under a peace agreement, but in January 2009 1,500 to 2,000 troops re-entered DRC under the banner of protecting Rwanda from Interahamwe.[13]

The big question was: what were they fighting over? Many have disputed Rwanda's claim that they were trying to prevent a second genocide by crushing the remnants of Interahamwe. Uganda's claims of national security and protecting people in DRC have also worn thin. The claims of various warring factions in DRC have also failed to win credence from the international community. Instead, the answer lies in the highly lucrative and global mineral trade. DRC is blessed, or cursed, with rich mineral deposits of copper, diamonds, gold and coltan – the last, as we have seen, an increasingly important strategic mineral for mobile technology. Fierce fighting has been partly paid for and sustained via mineral extraction. For example, Robert Mugabe and Laurent Kabila agreed that copper mining concessions could be used to pay for Zimbabwean military backing for Kabila's new government in the late 1990s. Furthermore, during the time that Uganda actively supported a rebel movement in DRC it became Africa's largest exporter of gold, even though it has no gold deposits of its own. Similarly Rwanda has become a major exporter of coltan, cassiterite (tin ore used in electronics) and diamonds; it is no coincidence that the mines are in areas that have been under the control of rebel movements linked to Rwanda. But this is not a strictly African problem.[14] The demand for these minerals comes from the wealthier parts of the world: Europe, North America and, increasingly, China. As with the wildlife trade itself, it is not poverty which drives the trade in minerals, but increasing wealth in the rich world. The materials for mobile technologies do not just appear from nowhere; they are mined in some of the world's poorest countries, and in areas rich in wildlife but riven by conflict.

Concern about the role of illicit mining in fuelling warfare in DRC prompted the UN to investigate; their 2001 report was damning, and pointed the finger at Rwanda and Uganda, but also criticized international buyers and even the Belgian national carrier, Sabena, for transporting minerals from the region. The UN panel of experts concluded:

> The consequence of illegal exploitation has been twofold: (a) massive availability of financial resources for the Rwandan Patriotic Army, and the individual enrichment of top Ugandan military commanders and civilians; (b) the emergence of illegal networks headed either by top military officers or businessmen. These two elements form the basis of the link between the exploitation of natural resources and the continuation of the conflict'.[15]

More recently, an investigation by the BBC in mid-2009 revealed how middlemen, or *negociants*, are able to source minerals from rebel-controlled mines in eastern DRC. Buyers in South Kivu showed little concern for whether the minerals have been mined illegally or if they were being sold to fund further conflict. Then the minerals are sold on to companies, like Thaisarco, a subsidiary of the UK-based Amalgamated Metal Corporation. Neither company is breaking any law, and they claim that they undertake due diligence to verify that the minerals are not mined in conflict areas under rebel control. But in a region like eastern DRC it can be difficult to determine where minerals have actually come from.

The problem is that the minerals can be accessed without the need of heavy machinery and large-scale companies.[16] One of the reasons the Botswana government has been able to benefit so much from its diamond deposits is because they are deep underground and can only be reached with large-scale mining infrastructure, so it is much easier to control their extraction and sale. In fact if you travel by bus from the capital, Gaborone, to Maun, the bus passes Orapa diamond mine, which has heavy security and clear fortifications. This is not the case in DRC. Mining is carried out by individual diggers with spades, who are vulnerable to militias who can demand a cut of what they find, or, worse, organize work gangs and take control of the mines. It is not just rebels that are responsible for violence around mining in DRC either. The national army has also been implicated in corruptly benefiting from the mineral trade, killing civilians, in torture and rapes during offensives designed to regain control of lucrative mines for the government. For example, parts of the national army formed a *mai mai* (self-defence unit) responsible for killing forty-four people in an operation to gain control of a cassiterite mine in Bisie, North Kivu in August 2009.[17] Part of the problem is that the army has been through a process called

5.3 UNHCR Nyakabanda camp for refugees from DRC (2007)

'mixage' whereby former rebel soldiers have been integrated into the national armed forces. It is impossible to separate out soldiers and rebels, with various factions joining and then leaving the army over the course of the conflict.

Turning back to wildlife, the ongoing conflict, especially in North and South Kivu provinces, is what vexes conservationists. The areas are important wildlife habitats and contain some of the world's most endangered and charismatic species. The conflict over access to the lucrative minerals directly harms poor communities and the world's last remaining gorillas. Local people in central Africa get the blame from conservationists but they are just as much under threat from the conflict as gorillas: they have to live with violence, poverty and disease produced by one of the world's deadliest wars. It is impossible to understand the threats to wildlife without understanding why conflict continues and why mineral extraction is so damaging.

Over the past fifteen years the public has become more aware of the problem of blood diamonds, or conflict diamonds. These were diamonds that were mined and traded (often illegally) and fuelled conflicts in places like Sierra Leone, Ivory Coast, Angola and the DRC. The NGO Global Witness launched a campaign in 1998 called 'Combating Conflict Diamonds' to focus international attention.[18] They are also the subject of Hollywood films, like *Blood Diamond*, starring Leonardo Di Caprio and Djimon Hounsou, which dramatized the production of blood diamonds in Sierra Leone and raised public awareness of the issue. Amid public pressure, the diamond industry and international community had to respond. In 2000 Southern African diamond-producing states met in Kimberley, South Africa, and by 2002 the UN had agreed to a resolution to create an international certification scheme for rough diamonds. In 2003 the Kimberley Process Certification Scheme, which was intended to clean up the trade, came into force.

> The Kimberley Process (KP) is a joint government, industry and civil society initiative to stem the flow of conflict diamonds – rough diamonds used by rebel movements to finance wars against legitimate governments.... The Kimberley Process Certification Scheme (KPCS) imposes extensive requirements on its members to enable them to certify shipments of rough diamonds as 'conflict-free'.[19]

The Kimberley Process now includes forty-nine members and covers seventy-five countries, and there has been a tendency to assume that it has 'dealt with' the problem. Certainly the reduction of conflict in Sierra Leone and Angola, the development of a market for 'ethically sourced' diamonds and efforts by large diamond companies like De Beers have reduced the flow of conflict diamonds. But the issue is not dead. DRC remains a source of conflict diamonds, and this is not a problem which is confined to that country. The diamond mining certainly fuels conflict over access to and control over the diamond fields, and it pays for the militias to continue fighting in the region. However, this would not be possible without a network of willing traffickers and a whole raft of external buyers who are unconcerned about the source of the stones. The award-winning Sierra Leonean journalist Sorious Samura made a special film as an extra to the DVD of the film *Blood Diamond*. It shows how he was able to pose as an illicit diamond dealer, claiming that he was selling stones from DRC; he found numerous willing buyers in the diamond district of New York – unconcerned about the source of the diamonds and their lack of certification under the Kimberley Process. These scenes are repeated across the major diamond-buying centres of the world, and underline the ways that illegal mining in DRC is heavily interlinked with global trading networks. It is not something that is bounded by the DRC itself, and neither is the long-term future of gorillas.

These complex dynamics of conflict were deemed to be one of the major problems that faced the world after the collapse of the USSR. With Cold War competition over, this seemed to be an opportunity for peace because wars would no longer be funded by superpowers engaged in proxy wars across the world. But instead different rebel groups and corrupt governments turned to international markets as a new source of finance for conflict. Analysts needed to explain this shift and so the idea of 'new wars' rose to prominence. It was claimed that the new wars of the 1990s onwards differed from the old wars which were characterized by distinct 'sides'; they had clearly defined professional armies and were subject to international rules of engagement. The new wars presented something different: they were characterized by multiple warring factions, not by distinct sides; they seemed to be ethnically rather than nationally based; they were driven and sustained by economic interests marked by criminal trading networks; and it was harder to distinguish civilians from combatants – embodied in the problem of child soldiers.[20]

Africa was the site of many of these so-called new wars: the Democratic Republic of Congo became the classic example. But for many people in Africa these wars did not seem all that new. The old colonial wars of suppression had not distinguished between civilians and combatants, and they were driven by the desire to gain access to lucrative resources like gold, ivory and slaves. In Angola and Mozambique, the proxy wars of superpower competition also had economic motives: rebel groups like UNITA were funded by access to diamond mines and the international markets. The claims that the new wars were ethnically based, more pejoratively referred to as 'tribal hatreds', do not stand up to scrutiny. But still the Rwandan genocide, and wars in Sierra Leone, Liberia and the DRC were all described as 'new wars', a convenient label for Africa's messy and seemingly intractable conflicts.

MADAGASCAR: LEMURS AND SAPPHIRES

The relationships between mining and wildlife conservation are not confined to gorillas in the DRC. Lemurs are also one of the world's iconic species, and they too are threatened by mining. I first went to the illegal sapphire mining area known as Ilakaka in 2001. I had always wanted to go to Madagascar to see its unique wildlife and strange plant life. As Madagascar is an island, many of the plant and animal species have evolved in isolation, and so it has a high rate of 'endemics' – species that are not found anywhere else in the world. As a result Madagascar is often described as 'megadiverse' because it is highly biodiverse and many species are unique to the country.[21]

Since the late 1990s it has also been identified as a mineral-rich country. It contains a wide variety of minerals including gold, nickel, bauxite, iron, chromite, titanium dioxide, beryl and quartz. The major investors in mining in Madagascar include QIT Minerals Madagascar (a division of Rio Tinto), which is engaged in titanium dioxide mining in southern Madagascar, and Canadian company, Dynatec, which has invested in a nickel mine in Ambatovy.[22] Until 2003 it was thought that Madagascar had no diamonds, but in that year a mining company with an exploration permit for areas in south-central Madagascar found two diamonds of 23 carats and 9 carats each. Since then, one Canadian mining company, Majescor, has been attracted to Madagascar, and has joined with De Beers to begin diamond exploration in north-western Madagascar. This has sparked investment in exploration by two other Canadian companies, Pan African Mining and

Diamond Fields International.[23] However, so far it is Madagascar's sapphires and rubies which have become the most important in terms of being major sources of revenue: the stones have been judged as the highest quality, and as global demand for coloured gemstones increases they could be a growing source of revenue to the government and private mining companies. If it were properly regulated then the mining sector has the potential to make Madagascar a very rich country.

Up until 2002, the international community had little faith in the government of Madagascar under its then leader Didier Ratsiraka, who was widely regarded as running a corrupt admin-istration and of overseeing rising levels of poverty and under-development. In the 2001 elections a new figure arose, Marc Ravalomanana, a businessman and mayor of the capital city, Antananarivo. Ravalomanana finally took over in 2002 after protracted disputes over who had won the election.[24] He appealed to the international community, and especially to the World Bank; he has a background as a businessman who began by selling yogurts from the back of his bike, a business that grew into a huge dairy products industry called Tiko. In 2003 the World Bank loaned the new Ravalomanana government US$32 million to gain control of its mining under a new five-year Mineral Resources Governance Project.[25] This was an attempt to get ahead of gem rushes and stop illegal mining from starting up; the idea was that the government would be able to control the location and devel-opment of mining to raise revenue for the country as a whole. As part of the project USAID produced a geological map of the country, identifying the potentially rich mineral deposits, for which the government could then sell licences for mining companies to exploit.

But gem rushes do not work like that, and the project turned a blind eye to existing illegal gemfields so it could concentrate on controlling future minefields. However, in gemstone mining the problem is the same as in DRC: the gems are in alluvial deposits, which means they can be dug up with a spade, and do not need lots of machinery and equipment that goes with deep-shaft mining. This has produced 'gem rushes', whereby people from all over Madagascar are attracted to the mines with the hope of making their fortune. Yet these fortunes could not be made without an international market for gemstones, and illegally mined stones are smuggled out into the international gem markets. This means that Madagascar has lost millions of dollars to the black market. A senior member of the Mineral Resources Governance Project suggested to me that US$100 million in gems are smuggled out of Madagascar, a figure supported by a 1999 World Bank Study.[26]

The problem is that many of the most important mineral deposits are under critical habitats or even inside national parks. Isalo National Park is an important protected area in Madagascar. It covers 81,000 hectares and is the country's most important in terms of tourism revenue. It contains numerous different species of lemurs (a particular attraction is Verreaux's Sifaka, endemic to southern Madagascar), and it is home to rare bird species such as the Benson's Rock-thrush. But it is also near one of the largest illegal sapphire mining centres in Madagascar, Ilakaka. Ilakaka rapidly developed in the late 1990s, after the discovery of high-quality sapphires. It was estimated that the population of the town increased from 30 people to about 100,000 between 1998 and 2000.[27] A member of the Mineral Resources Governance Project, who had previously worked as a gem dealer in Ilakaka estimated that US$4 million worth of stones were changing hands every day when trading was at its peak in 2001.[28] By 2004 the gem mining areas covered 4000 square kilometres. More recently, as Ilakaka has become a very

5.4 Verreaux sifaka, Isalo National Park, Madagascar (2004)

large mining centre, new areas have been subject to exploration by more recent migrants to the area, especially around Sakahara and the WWF-funded Zombitse-Mahafaly National Park, just south of Ilakaka.

Conservation organizations working in the area have tended to place the blame on diggers as a threat to wildlife habitat. Sapphire mining promises opportunities of instant wealth to Malagasy people living in poverty, so many of the 3,000 artisanal gold miners from northern Madagascar migrated to the Ilakaka and Sakahara area when news of the gem fields reached them. Since the gemstones can be extracted by digging with a spade, the poorest sections of Malagasy society have been migrating to Ilakaka because of the promise, or hope, of instant riches. The diggers do not threaten rare lemurs because they do not care about the environment or because they are ignorant of biodiversity. In fact many Malagasy groups feel a strong cultural attachment to nature: it is seen as sustaining life and numerous species have a spiritual significance. It is easy to focus on the diggers who arrive in Ilakaka as 'the problem', since they can be identified as the target group to be removed to protect wildlife, creating a neat solution. However, this misses the real problem and will not save lemurs; the diggers are part of a global system that links Ilakaka to gem buyers halfway across the world because there is a market for coloured gemstones in richer communities.

Economically impoverished Malagasy people (mostly men) travel to the gem areas to seek employment; they either work alone as diggers or are organized through clandestine networks headed by members of the Malagasy elite (there were allegations that the family of the former President Ratsiraka were heavily involved in organizing illegal ruby and sapphire mining in Vatomandry and in Ilakaka).[29] The diggers then sell their unpolished stones to gem dealers (usually Thai, Sri Lankan or Indian, but also African, European and North American) who have established a gem-buying business in the sapphire areas. Finally, the gem buyers traffic the stones out of Madagascar through airports or by sea, with assistance from key individuals within relevant customs departments, government agencies and local businesses. After this point the gemstones are part of the recorded international economy, and miraculously become legal products that can be sold to gem exchanges, and ultimately to customers in jewellery stores across the world.[30]

On the face of it the customer is buying a legal gem; once the stones are set in necklaces, rings, bracelets and earrings the ways that they have been produced are completely invisible. Sapphires are not covered by the Kimberley Process either, which is aimed at conflict diamonds. As with minerals in the DRC, diggers earn a fraction of the profits made by international gem dealers in Ilakaka and global gem dealers in Europe and Asia. The diggers are at a real disadvantage. They bring the stones to gem dealers in Ilakaka for valuation, and all the power lies with these dealers who are qualified to judge whether a stone is worth thousands of dollars or if it has imperfections which make it worthless. The diggers have no such knowledge or skill, so they have to trust the dealers to give them a fair price, and there are many stories of diggers being ripped off. The Catholic Relief Services staff said the town was full of stories about sapphires the size of footballs, fortunes won and lost in a single day, and dealers who knowingly cheat diggers by telling them their stones are worthless.[31] A walk through Ilakaka instantly reveals how global this town is. Just looking at the names of the shops is very telling: Congo Gems, Colombo Gems, Sri Lanka Safir and Thai-Mada Co-operation. They indicate that Ilakaka and its diggers exist in a global economy: without those foreign buyers and demand for colourful jewellery in the rich world, Ilakaka would simply not exist.

The heavy security, high walls and barbed wire around the shops indicate that Ilakaka is a violent place. The police face real challenges in the area because it is an illegal settlement, based on an illegal trade. During my last trip in 2004, as I was driving

5.5 International sapphire dealers, Ilakaka, Madagascar (2004)

through Ilakaka there were about 200 people were gathered on the main road; there was a lot of shouting and pushing. My vehicle seemed to be surrounded by an angry group, and I feared a riot would break out any moment. Suddenly a police officer appeared in the middle of the road and flagged down our vehicle. He jumped in and asked for a ride back to his station in the next town. He seemed exasperated, and said it was like that every day – fights, riots, lawlessness, 'shooting, like Texas' he laughed, referring to cowboy movies. The staff of Catholic Relief Services, the only NGO working in the area, constantly faced threats of violence, and were sometimes recalled to the capital for their own safety. The dealers also hire minders and security guards to protect them from theft and violent attacks. As in other illicit mining settlements violence erupts in the bars at night as frustrations mix with alcohol. While some miners bring their families with them, most do not, so the gender balance is very much skewed towards men. The small number of women often run market stalls selling food, clothing or mining gear, or they have no option but to work as prostitutes.[32] Ilakaka is the stereotyped frontier town of the American Wild West. It is not a place to wander around by day or night asking questions about the mining and gem dealing. There is a real atmosphere of intimidation and violence just beneath the veneer of a vibrant main road lined with stalls selling everything from tomatoes to medicines to mining equipment.

Ilakaka is not a pleasant place to be, and I was warned against going there many times. But it is just a few kilometres up the Route Nationale 7 to Ranohira, one of the bases for the Isalo National Park. Ranohira could not be more different; its restaurants and bars are aimed at tourists and the brightly painted billboards advertise trekking, visiting local villages and searching for lemurs, not gemstones. It seems a world away from Ilakaka, but the two places impact on each other, and the illegal mining has upset tour operators, hoteliers and conservationists keen on saving one of the world's rarest animals, the sifaka lemur.

Madagascar regularly appears on lists of biodiversity hotspots, world's most threatened habitats and most endangered species. For conservation NGOs it is important to work there, and the country's wildlife also acts as a critical fundraiser – everyone, it seems, loves lemurs. Wildlife Conservation Society, one of the many NGOs that work there, was very concerned about the potential impact of gemstone mining on biodiverse forested areas. They therefore formed part of a lobbying campaign directed at the government, to identify and designate 'no-go' areas for mining. The campaign centres around the idea that forests have long-term and sustainable economic value as places where ecotourists will visit. So, they argued, conservation 'pays'.

The Madagascar Parks Department (Association Nationale pour la Gestion des Aires Protégées; ANGAP) and donors like USAID were also concerned about the impact of illegal, unregulated mining in the area. They feared it would spread inside park boundaries and would be impossible to control. In fact, one of the park directors had to be relocated to the head office in Antananarivo once a rumour started that the best sapphires were inside the park boundaries and that he knew where they were. On top of that, ANGAP park managers were worried that, once the area has been exhausted of gems, the landscape will be left with ugly and unmissable scars from the mining, which will affect its appeal to tourists. The mining does leave a strange alien landscape marked by huge land areas where the vegetation has been stripped out, and with mining comes settlement. Driving through the area south of Ilakaka in 2004, I saw what I thought were small chicken coops made of wicker and fabric. Closer inspection revealed that these were small 'tent' structures, where the most recent migrant miners lived. The landscape was pockmarked with small holes, a sign of exploratory diggings. Within a short period of time these temporary settlements become permanent with offices, shops, bars, homes, churches and mosques.

Before mining began, conservation NGOs had tried out some community-oriented schemes to help manage the park and its surroundings. Projects tapped into existing cultural and spiritual attachments to Isalo National Park in the hope that they would have a positive spin-off for wildlife. Apart from wildlife, Isalo also contains a number of important ancestral graves and has cultural significance for Bara people in the area, who were engaged in a community-based conservation initiative in partnership with ANGAP and with WWF-Madagascar. ANGAP had pursued a policy of attempting to bring local Bara people into managing the protected area. For example, Bara people were given priority for employment as tour guides and game rangers. Half of the gate fees from the National Park are disbursed to Bara communities, and then used for community development projects such as building schools and clinics. ANGAP also runs education programmes to raise awareness about their conservation programmes in the park, which have worked partly because they intersect with pre-existing local ideas about appropriate uses of sacred landscapes. In return, Bara communities had to provide assurances that they would contribute to conservation by not breaking rules in the park by grazing cattle, burning grasslands and hunting. But the scheme has been undermined by the illegal mining in the area. Bara people regard digging as sacrilegious in the area where their ancestral graves are. This has produced violent clashes between the miners and the Bara community, some of them deadly. As a result the army was called in to prevent further fighting.[33]

None of this would happen without demand for the gemstones in the rich world. The miners have little option; they come from impoverished areas of Madagascar and the promise of sapphire fortunes is understandably attractive. The conservation organizations and donors have tended to focus their attention on controlling the miners, instead of controlling the traffickers in Madagascar and the gem dealers outside the country. This means the miners are the ones defined as criminals, as the threat and as a problem to be solved. But stopping the miners working in one area will merely displace them to another. A potential solution lies in offering alternatives to miners so they can make ends meet and feed their families without risking their lives in the illegal mines; this then needs to be backed up with strategies to tackle illegal gem buyers and smugglers and raising public awareness about the real costs of producing sapphire earrings and necklaces.

The disproportionate focus on the miners misses the real problem: demand from the rich world.

CONCLUSION

If we blame local people for the problems facing gorillas and lemurs, we are pointing the finger at the wrong people. The threats to wildlife in central Africa and Madagascar are not caused by the local communities in the area, but by organized groups supported and driven by demand for minerals and gemstones in the wealthier communities around the world. The focus of conservation organizations tends to be on the 'front line', tackling encroachments on national parks by local communities. This is clear in the ways charcoal collectors and sapphire miners are blamed by conservation NGOs and park authorities for threatening wildlife. The focus on local communities as 'the problem' makes it an issue that is bounded by the local area, and so it can be fixed by focusing on the local area. This misses the real problem, and fails to deal with the threats to wildlife in a way which might actually save gorillas and lemurs. This approach to conservation treats the symptoms, not the cause.

Conservation NGOs are often aware of this, but when faced with taxing problems on the ground and lacking the tools to tackle the underlying problems they fall back on the tried and tested. They understand 'protected areas' as a conservation method, and promote a message that portrays local people as the problem. This is compounded by the ways they have to present and promote campaigns to raise funds and international support; so again they retreat to the tried and tested message: here is the problem as we see it, and the simple solution is to 'act now' by donating to the NGO. A clearer message about the complex and important role that local communities have to play in longer-term conservation is lost, as are the complexities of the international framework in which conservation problems are produced.

In order to understand why gorillas are threatened with extinction we need to understand the history of conflict in the region, the current power dynamics and the economic underpinning for continuing warfare. Once we understand why rebel groups are still so active in the central African region, and why they use protected areas to raise revenue from the charcoal trade and from mineral extraction, our approach to conservation might change. The factors that underpin and sustain conflict in central Africa need to be resolved before gorillas can be made secure. It is important to understand the demand side of the equation, because it is demand for coltan, cassiterite, diamonds and other minerals that provides the economic underpinning for continuing conflict. Conservation organizations need to tackle the demand of these minerals in the wealthy world, and make it clear to the public that they should demand 'gorilla-friendly' mobile phones and refuse to buy conflict diamonds.

This is equally the case for sapphire mining in Madagascar. It is easy to point the finger of blame at local diggers as the problem and as the threat to lemurs. However, they are only in Ilakaka because there is an international market for high-quality sapphires. Poverty in Madagascar makes the promise of sapphire wealth more appealing, but it does not cause and produce a phenomenon like the Ilakaka mines. Diggers will migrate to the area as long as there is demand for sapphires from Madagascar, and lemur habitats might be the unfortunate casualty. Conservation organizations need to educate consumers, and those customers need to demand that the sapphires they buy are produced under the highest ethical and environmental standards. The continuing failure to do this merely allows us to blame the world's poor for conservation problems created by the wider global context of demand for wildlife.

CHAPTER SIX

TOURIST SAVIOURS

TOURISTS ARE OFTEN PRESENTED as the potential saviours of wildlife, precisely because so many of us will pay to travel halfway across the world to see animals in their natural habitats. But the role of tourists is very complicated, and they can be involved in the illegal trade in endangered species either as willing participants or as unwitting traffickers.

The case of abalone in Australia is very revealing. I was once greeted there with the curious sight of a group of tourists carrying very large polystyrene cooler boxes back onto their tour bus. Curiosity got the better of me and I asked around; the answer lay in the grey area between legal and illegal wildlife trade. The coach was carrying an organized tour from China, and for many of the visitors a tidy profit could be made from buying boxes of dried or fresh abalone from supermarkets in Brisbane. Australia is one of the primary producers of legally fished abalone. As we saw in chapter one, the abalone trade is highly lucrative, and the demand is met from both legal and illegal sources. In this instance tourists may well be fuelling illegal fishing of abalone, but they believe they are buying a legal product that can be legally exported. The Australian authorities readily admit that illegal fishing is a real problem and they have found it hard to enforce fishing rules.[1]

Many of us are caught up in this grey area, but few of us really think about whether we should buy a souvenir made of shells in Sri Lanka or take advantage of the good prices for wooden carvings in Kenya. Our behaviour on holiday can affect wildlife in unexpected ways as well – in terms not just of souvenir buying, but of how we interact with animals on holiday. In many holiday destinations we see people wandering around with animals, encouraging us to have our photos taken with them – lorises, iguanas, gibbons, marmosets and even baby elephants are touted in the world's tourist playgrounds.

The quickest glance at holiday brochures shows us that wildlife tourism is big business, so the argument goes that wildlife will be saved if it is developed as a tourism product which attracts investment for sustainable development. However, this is based on a very crude link between tourism and wildlife conservation, which fails to recognize the potentially damaging impacts of tourism. For most tourists, souvenir shopping is an essential part of any holiday. In many holiday destinations, souvenirs are a memento of the local area and so they are produced from local materials such as wood, shells, skins and stones. As a result tourists can (often unknowingly) illegally export endangered species as souvenirs, especially trinkets made from turtle shell, seashells and some hardwoods, particularly ebony wood.

But the impact of tourism is much bigger than that, and is related to the links between tourism and organized crime. One of the challenges facing the tourism industry is the ways in which hotel construction can be used to launder money from organized crime. This is important because the linkages between 'legitimate business', illicit networks and state agencies can mean that conservation issues are pushed aside by more powerful interests in favour of building hotels, access roads, restaurants and bars. In this chapter I will explore the connections between tourism, criminal networks and conservation to explain why tourism is not necessarily the answer to saving the world's endangered wildlife.

LEAVE ONLY FOOTPRINTS? SHELL COLLECTING AND SOUVENIR HUNTING

While conservation NGOs, governments, and tour operators try to convince us that tourism can save wildlife, our holidaying habits can have a really damaging effect on wildlife. We are sold the idea that tourists will pay to see wildlife, mountains, forests, coral reefs and beaches, and this gives local people and governments an incentive to conserve them. But this is an oversimplification. Tourists, and the infrastructure that comes with them, can have a devastating impact on wildlife. For example, some tourists use the wildlife trade to fund their trips. In 2009 police and customs officers seized two tortoises they believed were indigenous to Corfu. The detective in charge of the case pointed out that there was a real problem with people taking 'one or two' animals from their chosen holiday destination, to be used as family pets or to be sold to cover the costs of the holiday. Even though this kind of smuggling is small-scale compared with organized trafficking networks, it still has an impact on wildlife.[2]

Problems do not arise just from the impact of tourists deliberately collecting endangered species for economic gain; they are also produced by everyday ad hoc choices made by tourists. Tourist demand for seashells is a good example. Most of us at one time or another will have picked up a pretty shell from the beach as a reminder of a day out or of a holiday. It feels as if shells are an endless renewable resource – they wash up on seashores all around the world. Shells are also made into souvenirs and jewellery for tourist consumption. But the removal of shells negatively impacts on the marine environment.

In the Maldives, one resort tried to bring home the message to tourists that taking shells, even small ones, from the beach was harming the very marine environment they had come to enjoy. Because tourists were picking up shells from the beach, the local hermit crabs were finding it difficult to find a suitable home. As they grow, the soft hermit crabs look for empty shells. Tourists were removing so many empty shells that the hermit crabs turned to anything that they could crawl into for protection. This included debris that washed up on shore, producing the strange sight of scuttling pen lids and dials from radios. The problems for the hermit crabs were caused by the everyday individual actions of the thousands of tourists who visit the Maldives every year. The apparently small individual choice to pick up a shell from the beach had a cumulative effect on the marine environment, an effect that would be invisible to tourists because their stays are so short.

In Sri Lanka it is common for beach hawkers to sell large colourful seashells to tourists, and shops are full of souvenirs and jewellery made from shells. A quick glance round the tourist shops and stalls reveals a lively trade in curios made from corals, seashells and dried animals (pufferfish, starfish and seahorses). Hotels, spas and restaurants are also decorated with marine life – shells, dried starfish and beach pebbles create the relaxing 'global zen' mood of the swanky beach resort.[3]

Turtle shell is perhaps one of the most controversial products for tourist consumption. The demand for tortoiseshell products is a critical factor in the decline of turtle populations globally, along with egg collection and hunting turtles for meat. The

6.1 Poster highlighting the plight of homeless hermit crabs (2007)

hawksbill turtle has seen an 80 per cent decline in its population and is listed under Appendix I in CITES, which constitutes a total trade ban. However, tortoiseshell remains an important product in international trade, and the tourism industry is one of the main sources of demand. In 2009, TRAFFIC reported that turtleshell products were being offered to tourists in Papua New Guinea. The survey found that a total of 1,441 marine turtle and 12 freshwater turtle products (mostly jewellery items) were on sale at the airport and duty-free shops around the capital, Port Moresby. Even more alarming was the fact that 99 per cent of sea turtle products were made from hawksbill turtles. It is easy to point the finger at the shops selling the turtle products, but they would only sell them if there was a demand for them from tourists themselves. Papua New Guinea has a long history of using turtleshell for earrings, decorative motifs and spatulas for taking lime while chewing betel nut. However, this use has been small-scale and had a minimal impact on local turtle populations, while the development of turtle-based souvenirs for tourists presents a much greater threat.[4]

In 2007 WWF urged British holidaymakers to be more careful about their souvenir hunting, and encouraged tourists to check whether handbags, belts and jewellery were made from endangered species. In 2006–7 there were 429 seizures of illegally imported wildlife products, up from the previous year's total of 302. The main items were snakeskin goods (mostly handbags and shoes), elephant ivory carvings and more than half a ton (605 kg) of TCM. Customs officers also identified coral as a major problem; tourists who bought coral jewellery and trinkets were often unaware that they were governed by CITES legislation and it was illegal to bring them back to the UK.[5]

Other countries have also tried to educate tourists. In 2008, the Belgian government launched a campaign with TRAFFIC and WWF-International to educate tourists about the potentially harmful effects of their souvenir buying. The campaign was called 'Leave a future for your souvenir'; leaflets and posters were displayed in Brussels airport and given out by tour operators. Belgian tourists mostly brought in illegal items from Thailand, DRC and Senegal, including bracelets made from elephant hair (which are widely available across Africa) and jewellery made from coral and ivory.[6] TRAFFIC produced a series of online videos to educate tourists, including one called 'Shop Carefully' which encourages tourists in the East Asian region to think carefully before they buy, and recommends that if they are in doubt they should not buy. WWF and TRAFFIC have also produced stickers, badges, posters and leaflets to educate tourists about the harmful impacts of souvenir hunting.[7]

6.2 Shell souvenirs for tourists, Thailand (2008)

It is not all bad news. Educating tourists about the impacts of their choice is certainly one way of tackling the problem. On the other side of the world, in the Dominican Republic a government crackdown on sale of hawksbill shell souvenirs to tourists resulted in a 99 per cent drop in items for sale between 2006 and 2009. The crackdown coincided with campaigns to encourage the use of cow horn in place of tortoiseshell and attempts to educate tourists about the impact of buying the products. The successes in the Dominican Republic have given conservationists some hope that tourist attitudes can be changed and sellers can also change their habits. After all, if there is no demand, there will be no trade.[8] It is important for tourists to think twice before they buy souvenirs made from coral, tortoiseshell or seashells because it has such a negative impact on the very places and wildlife they have come to see.

THE WOODEN CURIO TRADE

It is not just marine life that is in demand for tourist souvenirs. Wooden products, everything from chairs to carved animals, are also an extremely popular choice as souvenirs. Tourists are often delighted to see the curio stalls and markets that dot roadsides and spring up in cities all over Africa. Buying a wooden carving of an animal, a bowl, a chair or a figurine is a key part of the holiday experience. It is common to see tourists on flights out of Johannesburg boarding planes with enormous wooden giraffes wrapped up with paper and string. TRAFFIC identified the wooden curio market as an important driver of deforestation and unsustainable harvesting of valuable hardwoods in Africa. In 2002 and 2003 South Africa imported approximately US$400,000 worth of curio products from Malawi, and this only represents the formal and legal trade; many more carvings arrive via illegal trading routes.[9]

Rosewood and ebony wood are much sought after and the trade in them can be highly lucrative. Official figures from the government of Madagascar show that the export of forestry products including rosewood, ebony wood, pinewood and medicinal plants earned the country an average of US$6.5 million per year from 2001 to 2007. A CITES review of wildlife trade in Madagascar found that while there was a greater degree of commitment to developing and enforcing CITES regulations under the Ravalomanana government from 2003, different stakeholders (such as wildlife traders, relevant ministries and NGOs) had a limited understanding of CITES regulations and there was a lack of political will at the national government level.[10] The trades in rosewood and ebony wood are not banned under CITES; however, there are concerns that these valuable hardwoods were being harvested in illegal and unsustainable ways.

In Madagascar there is a curio market aimed at tourists, situated halfway between the airport and the centre of the capital city, Antananarivo. Wooden souvenirs are highly popular and range from small carved lemurs to the full-size African-style chairs consisting of two pieces of wood are cleverly slotted together. The carvings on all these products are intricate and demonstrate incredible skill. Wooden inlay is also very popular, with patterns and pictures are created on furniture, boxes and other trinkets through the use of different coloured woods.

Sellers are usually keen to point out that the sturdier items of furniture are made from rosewood because this is a very durable tropical hardwood that takes many decades to reach maturity. Apart from furniture and souvenirs in Madagascar it is also in demand for making musical instruments, such as guitars, across the world. The market stall operators could not understand why I was refusing to buy rosewood products, and even worse, why I was asking for curios made out of apparently inferior materials. They took offence, believing my requests were motivated by a lack of trust: that I did not believe their claims that the products were made from rosewood. Most tourists visiting Madagascar would be unaware of the problems associated with buying wooden curios, especially rosewood. During 2009, the change in government in Madagascar increased concerns about the impact of rosewood harvesting. In 2009 the new Rajoelina government banned the harvesting of all precious woods, but then gave permission for containers filled with illegally harvested rosewood to be exported. This prompted a consortium of global NGOs including Conservation International, the Madagascar Fauna Group and the Durrell Wildlife Conservation Trust to call for an international boycott of Malagasy rosewood.[11]

Ebony has a clearer public profile, but there is still considerable confusion over whether it is legal to buy ebony products and bring them back home. Again I draw on experience from Malawi in the mid-1990s. Many of the backpackers travelling the route through Central Africa down to South African passed through Malawi – Cape Maclear on Lake Malawi and the mountains around Zomba were popular stop-offs before travelling through the Tete corridor in Mozambique to Zimbabwe or on to Zambia. Enterprising curio dealers had come up with a solution to the problem that backpackers did not want to carry the intricately carved wooden 'Malawi chairs': they offered to package and post them. Some of the curios were made from ebony wood, which is restricted under CITES. Clearly other tourists had asked about this because there was some awareness that ebony was an endangered hardwood. When I asked the curio sellers about this they would immediately respond with: there are two types of ebony, ebony one and ebony two; ebony one is banned, but you can take ebony two back to the UK with no problem. This sounded convincing, but hard to check in Lilongwe in the mid-1990s before internet cafes became as ubiquitous as they are now. In fact it is not clear at all when it is legal to export ebony products; and even where the product is legally exported the wood itself may have been illegally and unsustainably logged.

Ebony is a very valuable hardwood; it is so dark that it looks black, and so dense that once polished it can look like stone. There are a number of tree species across the world from the *Diospyros* group that yield ebony. The African blackwood tree was once classified as ebony and is still routinely referred to as African ebony; it is on the endangered species list. It can take decades for the trees to mature, up to eighty years to reach 40 centimetres in height, which means they are vulnerable to unsustainable harvesting. It is used in numerous products from earrings to cigar boxes and musical instruments. The best known are piano keys, but it is also much in demand across the world for high-quality bagpipes, oboes and clarinets. Indeed, in 2007 bagpipe owners were encouraged by Flora and Fauna International to contribute to an African blackwood replanting scheme to offset their environmental impact.[12]

Flora and Fauna International have been engaged in developing community forestry schemes in Tanzania to try to ensure communities receive some of the benefits of blackwood logging in their area.[13] Understandably, the vibrant carving industry in Mozambique, Kenya and Tanzania has vehemently opposed any attempt to ban or restrict international trade in African blackwood. It is estimated that 80,000 carvers rely on the industry in Kenya alone.[14] Moves to get ebony and African blackwood listed as CITES Appendix II in the mid-1990s failed amid protests from Tanzania and Mozambique. African blackwood is clearly part of a massive industry which is sustained by global demand for carvings, furniture and musical instruments. It is another example of how legal and illegal harvesting and trading are interlinked and very difficult to separate out. It may be legal for tourists to bring a rosewood souvenir back from Madagascar or an ebony wood mask from Malawi but the demand for such curios can drive unsustainable harvesting of valuable hardwoods. Any attempt to deal with the problem needs to tackle tourist demand as well as providing alternatives for local communities who rely on wood carving. A simple blanket ban on wood carvings will not work.

Deforestation as a result of the curio trade is a real concern for many conservationists. However, the curio industry is important to many communities across Africa, especially in tourist areas. There is little to be gained from trying to stop people selling wooden crafts by the roadside if they have few alternatives for earning an income. The stalls and markets would not exist unless there was a demand from wealthy tourists visiting the country. Therefore we cannot be complacent about the idea that tourism is necessarily

6.3 Wooden carvings for tourists, Cape Verde (2010)

good for wildlife or the wider environment; instead tourists can be involved in driving deforestation in poorer countries. In the case of wooden curios it is again demand from the wealthy world that poses a problem rather than poverty in the developing world which drives deforestation. Conservationists need to focus their attention on educating tourists and offering alternatives to poorer communities rather than simply trying to ban the trade in wooden curios or shutting down roadside markets.

THE RESTAURANT TRADE

Tourists do not just consume wildlife as souvenirs. For holiday-makers one of the greatest pleasures is trying out the local cuisine. In beach resorts this means there is a huge demand for seafood, especially lobster and conch, as part of the holiday experience. But we tend not to think very carefully about how the food we are eating is produced and whether it has a negative impact on the very environments we have come to enjoy on our vacations. It is not uncommon to see local fisherfolk out looking for lobster in holiday destinations around the world. As boatloads full of scuba divers and snorkellers head out to particularly interesting patches of coral reef, they pass local fishing boats with people 'free diving' for lobsters on the reefs. The fishing is there to meet the demand from tourists in local restaurants.

When I was working in Belize in the late 1990s, one of the hottest topics was how to stop unsustainable and illegal harvesting of lobster and conch. The fears among the conservation community in Belize were expressed in the Lamanai Declaration in 1997,

6.4 Seafood restaurant, Phuket, Thailand (2008)

which specifically detailed lobster fishing as a major threat to the environment. Lobster was a highly lucrative commodity, netting the country US$8.8 million per annum in the mid-1990s.[15] While most of the lobster caught in Belizean waters is for export to North America and East Asia, a proportion of it ends up in tourist restaurants. Lobster fishing is regulated by a system of permits allocated by the Government Fisheries Department. Apart from the permit system, the Fisheries Act of 1948 also restricted what kinds of lobster could be caught; it set a legal size limit to prevent exploitation of juvenile animals (lobster with a carapace of less than 7.6 cm), introduced a closed season from February to June for lobster fishing and banned the removal of berried females (those carrying eggs) those with soft shells (because they are in mid-moult), and the use of scuba equipment while fishing.[16] However, this proved to be very difficult to enforce and conservation organizations persistently complained that corrupt individuals in the Fisheries Department were allocating permits unfairly.[17]

For example, A&J Aquaculture, owned by James and Andrew Wang, was given a licence to fish for undersized lobsters by the then Minister of Agriculture and Fisheries. The licence was issued because the company wanted to start up a lobster farm, but it had failed to raise adult lobsters from larvae. So the company was allowed to catch juvenile lobsters to start their farm. The Department of the Environment claimed that this was in contravention of Fisheries Department legislation, and was reported to be very unhappy with the decision.[18] One newspaper reported that the decision to grant the licence to the Wang brothers came just three days after the Taiwanese Foreign Minister had approved a loan of US$20 million to the then UDP government to develop their southern highway development project.[19] Clearly, there was a strong local perception that James Wang, who had a history of corrupt business activities, was being allowed to flout the national regulations on lobster fishing.

The Fisheries Department was also criticized for failing to send inspectors to restaurants to check that they were not using lobster out of season or using baby lobsters and conch.[20] The Conservation Compliance Unit of the Fisheries Department was responsible for monitoring fishing, but had its budget cut so that its enforcement and management capabilities were stretched beyond capacity.[21] Conservation organizations appealed to the demand side of the equation rather than just focusing on those fishing for seafood; they tried to encourage tourists to order only whole lobster rather than salads that used chopped lobster or lobster tails. It is more difficult to 'hide' undersized and juvenile lobster that has been illegally fished when it is served whole – once it is chopped up in salads it is impossible for consumers to tell whether the animal was a legal size.

However, the relationships between tourism and the global trade in lobsters are even more complicated than the journey from the sea to a restaurant table. Bernard Nietschmann's study of the relationship between cocaine, scuba equipment and illegal lobster trading in the Caribbean is instructive here. It shows that the demand for lobster in the rich world means that it is interlinked with other products that are illegally traded. Nietschmann's study demonstrates that illegally fished lobster and cocaine is traded via small fishing boats from the Caribbean coast of Nicaragua, up through Belizean waters, along the cost of Mexico and into southern Florida. The people who run the small fishing boats that transport the lobsters and drugs obviously do not want to make a return journey with empty boats. So they buy up scuba equipment and transport that back down the coast, selling it in the booming scuba tourism industry.[22] As with the undersized lobsters eaten by tourists, the demand for this comes from the rich world but the supply comes from the poor world.

SHOOTING WILDLIFE

The dynamics of wealth and poverty also explains another phenomenon in tourist resorts which many of us are familiar with: the use of animals for photographs. Playa Del Carmen on the Caribbean coast of Mexico is a booming and fast-growing tourist resort. In the evenings tourists enjoy a stroll down the central avenue, La Quinta, which is lined with restaurants, shops, stalls and bars. You can also see people offering small animals for tourists to have their photos taken with them. On a recent trip, I was intrigued by someone who claimed that the proceeds from a photo with a marmoset would contribute to marmoset conservation and breeding programmes. On closer inspection it became clear that this was a promotional activity by a well-known photo shop, since the photos had to be developed on site.

This is one end of the spectrum; at the other are animals trapped in the wild – any sharp teeth and claws are removed and then they are drugged to make them more docile and amenable to constant handling by tourists and photo hawkers. The Bangla Road in Patong on Phuket Island in Thailand is commonly known as a place filled with go-go bars and strip shows, but it is also a prime location for hawkers to sell the chance to have a photo taken with a cute animal. Young gibbons are a particular favourite, and they are often dressed up in nappies and baby clothing to make them even more appealing. However, tourists are not necessarily aware that the mother will have been shot to separate her from the baby. Such animals have a short life expectancy because they have been removed from their family group and the owners do not necessarily know how to care for

6.5 Marmoset offered for tourists to photograph, Playa del Carmen, Mexico (2009)

them. The same goes for slow lorises, a small nocturnal mammal with large appealing eyes. Lorises have very sharp teeth, and the photo hawkers remove the teeth to ensure the animals do not bite the tourists; but this affects their ability to feed properly, and is a painful procedure. Lorises can be drugged to make them more compliant to handling, and quieter as their owners walk through the streets trying to drum up business.[23]

It is not just small animals that are the problem: the ways elephants are used in tourist resorts in Thailand has also caused real concerns. Following the logging ban in 1989, the working elephants used in forestry plantations needed to find alternative forms of employment. Asian elephants need to eat around 120–200 kgs of food per day; and in Thailand each trained elephant comes with a mahout who either owns the elephant or works closely with the animal, making it safe to be around. For the mahout the elephant is a source of income as well, but obviously it costs a great deal of money to care for an elephant which may live with the mahout for fifty years. After having been made redundant from the logging industry, some of the trained elephants and mahouts tried to make a living by begging in cities and tourist resorts. This caused real problems – the elephants were a danger to pedestrians and traffic, they were scared by the lights of urban areas and walking on tarmac roads hurt their feet.[24]

Over the past twenty years, many of the trained elephants have been funnelled in to tourist camps where tourists are offered rides, or used in shows where they display their traditional logging skills, or tourists can pay to wash, feed and care for them. However, even though it is against the law in Thailand, there are still some elephants turning up in tourist resorts, where mahouts use them to beg or offer the chance to pay to feed the elephants or have your photo taken with them. Obviously baby elephants are the ones that tend to generate the greatest interest from tourists, who are attracted by the opportunity to interact with them.

It is easy to criticize the mahouts who contribute to the phenomenon of 'street-wandering' elephants. They are often cast as cruel people who aware that they are acting illegally, know that the elephants are in pain from walking on hard roads and are scared by lights when they are out at night. However in 2008 I was working in Thailand, researching the different ways in which elephants were used in the tourism industry, and it became clear that this was an oversimplified stereotype. Mahouts were aware that the practice was illegal and potentially distressing to the elephant. But they also expressed how much they cared for the animals and explained that often there simply did not seem to be any other option. They needed to find a way to make a living, which included keeping the elephant they had worked with for decades, and pointed out that unless there was a demand for this from tourists then they would not bring the elephants into towns and resorts.

The Thai Elephant Conservation Centre and some of the privately owned elephant camps have tried to explore alternative forms of employment in more regulated elephant camps which offer tourist experiences but which also manufacture elephant products, like elephant dung paper and fertiliser.[25] Tourists need to be more aware of how their choices affect animal welfare. In Thailand there are numerous well-regulated camps where tourists can interact with elephants that are well cared for and where mahouts have good working conditions. Tourists and tour operators need to ensure that they use these well-maintained camps instead of contributing to an illegal economy of elephants being used for begging and working in poor conditions.

BUILDING TOURIST RESORTS

It is not just the ways tourist collect shells, buy souvenirs or have their photos taken with wildlife that affect the environment. When we are on holiday we do not really concern ourselves with how the hotel was built, whether the beach is a 'natural feature' or how the sewage system works for local restaurants. Tourism is promoted by a whole range of powerful interest groups, from the World Bank to national governments to tour operators and conservation NGOs as a possible solution to the problem of sustainable development. It is assumed that because tourists are attracted by the natural world then it makes sense that tourism can be a way of generating income from protecting the environment. Tourists are keen to see apparently pristine beaches, experience safaris in national parks or go on adventurous rafting trips down untamed rivers. Conservation NGOs claim that it pays to conserve the 'natural' environment because tourists will come in their droves to see and experience it.

This is an especially powerful argument for developing countries. Their seeming lack of development has been reconfigured into a virtue – the lack of industrialization and urbanization means that their beaches, forests and mountains savannahs are a global commodity for sale on the international tourism market. For conservation NGOs, developing countries offer a chance to experience wilderness on holiday. But this fails to acknowledge the ways in which the key natural attractions for tourists have been transformed and produced by the tourism industry itself. When criticisms are levelled at mainstream or mass tourism as bad for the environment, conservation organizations and tour operators point to ecotourism as the answer. The precise definition of ecotourism is still a subject of much debate; however, it is clear that it relies on the idea that places and cultures are pristine, unspoiled and untouched by Westernization, industrialization, and even mass tourism.[26] Contrary to what we might think, wildlife or nature-based ecotourism redefines, presents and can even *create* areas of 'wilderness', which are, more often than not, based on notions of people-free landscapes.[27] Ecotourism developments are not 'impact-free' and can be linked with criminal activity in surprising and environmentally damaging ways.

There is no doubt that Western tourists love beaches. Tropical holiday brochures encourage us to experience perfect relaxing beaches, which appeal to a Western taste for the exotic and unspoiled 'getting away from it all' experience. This means that any land near good beaches is highly sought after and is very valuable. This love of beaches can produce a temptation to build artificial beaches or increase the size of existing beaches to meet tourist expectations. Few of us realise that the beaches we sunbathe on might be artificial structures, created by the tourism industry itself.

Playa Del Carmen is one of Mexico's tourism hotspots. It is one of the fastest growing urban areas anywhere in the world, and it is all built on global tourism. However, the wide sandy beaches at the front of the four- and five-star resorts in Playa Del Carmen are not quite what they seem. While tourists soak up the sun, the beaches are being slowly and carefully 'built'. This is done in two ways. The first is to place large sandbags – locally known as 'whales' – at the shoreline. The whales prevent the sand from washing back into the sea and help to build up sandbanks, and thereby expand the width of the beach. Guests are told that the whales protect the beach front from storm damage, which they may well do, but their primary purpose is to build up sand for guests to enjoy. The second method used is to pump sand directly from the sea floor up on to the beach to build it up. Large green hoses lead from the sea up to the top of the beach, pouring out a mixture of sand and sea water. If tourists looked a few hundred metres down the beach to an area not owned by a hotel, they would quickly see what a more 'natural' beach would look like; they are much narrower and dotted limestone outcrops – not exactly the picture-perfect beach for sunbathing and building sandcastles. Beach building in this way redesigns and produces a particular environment for tourist consumption. While tourists may think they are looking at a pristine tropical beach, they are in fact experiencing a human creation.

Beach building produces attractive features but it is highly damaging to the marine environment. Removal of sand from the seabed changes the currents in the water, which can have a negative impact on coral reefs and marine life further away from the beach. It also means that if a hurricane or major storm hits the coast, the extra sand is carried out to sea and then smothers coral reefs, making it even harder for them to recover. Beaches are also the nesting places of one of our favourite animals, the turtles, and beach building disrupts traditional turtle-nesting sites. Further

6.6 Beach building, Playa del Carmen, Mexico (2009)

down the coast in Akumal, the beach is well known for being a place where loggerhead and green turtles feast on eel grass and nest on local beaches. The Centro Ecological in Akumal is keen to develop sustainable tourism in the region and turtles are a major attraction in the local area. Volunteer workers engage in education and communication activities for local residents and visitors: they give talks about turtles and have developed a display to explain the need for turtle conservation. In addition, volunteers from the Centro Ecological monitor tourist interactions with the turtles and mark out the nests on the beach to prevent disturbance. In 2009 they had ninety-nine loggerhead nests and ninety-two green turtle nests.[28] However, all along the Caribbean

6.7 Turtle nest marker, Akumal beach, Mexico (2009)

coasts of Mexico beaches and hotels are being built, with negative consequences for turtle nesting sites.

The controversies surrounding beach building are hidden from the tourists who come to enjoy them. The case of the Caye Chapel resort in Belize is a good example of the ways in which the environment can be completely remodelled to appeal to tourist tastes and then be promoted and sold as a pristine tropical beach location.[29] Its owners claim, somewhat paradoxically, that it 'has been developed as a pristine oasis'.[30] In fact, the island was completely relandscaped and artificially expanded to provide an exclusive tourist resort. Caye Chapel describes itself as offering:

> stunning beachfront community living. Enhanced by the Caribbean Sea on its eastern shores, with a spectacular 18-hole, Par 72 championship golf course (USPGA rated)

along its western edge. Featuring a deep-water marina welcoming yachts up to 140 feet, private airstrip, and 25,000 square foot clubhouse with a magnificent custom designed bar and restaurant, conference facilities, large swimming pool complex with bar, and over 2 miles of pristine sandy beaches. The entire property has been developed at the highest possible standards and is meticulously maintained.[31]

The development of the island included dredging for sand to build the beach and to allow larger boats to dock at the island. In the process, the dredging stirred up the seabed, disturbed lobster fisheries and destroyed Caye Caulker fishing grounds.[32] This reconfiguration of the environment for tourism development also led to an illicit trade in sand. The sand pirates operated at night, removing sand from neighbouring islands, which was then used to build up artificial but aesthetically pleasing beaches on sand-free coral atolls.[33] This phenomenon has been created because of the tourist desire to have destinations that conform to their stereotyped image of a pristine Caribbean paradise of turquoise water, white sand and coconut palms. Ambergris Caye has suffered from the activities of sand pirates more than other islands, with tons of sand being removed over time. The sand is an important part of the wider ecosystem on the island that supports coral reefs, mangroves and a variety of fauna, and so sand piracy has a significant negative impact on the environment.[34]

When I was there in the mid-1990s, Caye Chapel was something of an embarrassment. The speedboat that ran locals and tourists between Belize City, Ambergris Caye and Caye Caulker passed it numerous times per day, and the reconstruction of the environment was plain for all to see. From time to time the boat captains would point at it and tut with disapproval. They would make reference to the rich American who owned it, how he planned to fly tourists in directly from Miami for holidays centred around diving, gambling and golf. At the time Belize was promoting itself as an ecotourism destination, but the rebuilding of an entire island to create an ideal tourist paradise ran against this spirit. For some in Belize, Caye Chapel was a glaring example of what happened to the environment when a corrupt politician and a foreign millionaire got together; they felt that the development was all about money and there was no one willing to stop it.[35] Caye Chapel has proved to be a controversial tourist development that highlights such links between foreign investors, support from local political and economic elites and local perceptions of corrupt business practices. There was a great deal of local speculation about the way in which the owner of the island appeared to be able to dredge around the island and build a beach without proper permission. Those who opposed the development of Caye Chapel speculated that the owner was allowed to undertake such activities because he was protected by the highest political authorities in the country in return for free trips to the island, and that the owner had been careful to pay the relevant officials in order to avoid environmental regulations.[36]

One of the problems for anyone concerned with the environmental impact of tourism and the need for regulations to protect wildlife is links between tourism and criminal networks. The tourism industry has proved to be a place where illicit business interests can meet up with corrupt politicians, because the arrival of tourism is often associated with an increase in crime, prostitution and the supply of drugs. Drugs may be only one part of a broader criminal business that includes legitimate business on the one hand (such as tourism), and bank robberies, car theft, arms trafficking and kidnapping on the other.[37]

The demand and supply routes for drugs, particularly cocaine, had a significant impact on the direction and rates of development in the tourism industry in Belize. For example, it was reported that cocaine trafficking brought a new spurt of wealth to the local economy in Placencia, where the local press noted the appearance of new speedboats and the beginnings of a construction boom.[38] Similarly, there was local speculation by critics of the involvement of foreign interests and possibly criminal elements that entire resorts were bought with millions of dollars in cash, derived from the drug trade. In 1999 James Kavanagh, a citizen of Colorado, was expelled from Belize and escorted out of the country by US marshals and drugs enforcement administration officers. He had been a resort owner in Cayo District, but had been found to be engaged in money laundering and the drugs trade, using the resort as a legitimate business front for illegal activities.[39]

The existence of illicit links between the drugs trade and tourism development means that enforcement of environmental legislation is problematic in Belize. The country has an extensive framework of environmental legislation, including the Environmental Protection Act, and a Department of the Environment. A common complaint among locally based tour guides and the conservation community was that while legislation was in place, it was rarely enforced. Godsman Ellis, President of the Belize Ecotourism Association, stated that policing was not carried out in the face of the country's financial resources, limited workforce, inadequate technological and administrative resources, and its open borders; he suggested that very few cases of environmental violations had been brought before the courts despite the legislation that had been put in place.[40] One interviewee, who was involved in marine conservation in Belize, stated that although environmental impact assessments were legally required for all new tourism developments, the government office responsible for monitoring such assessments was completely overwhelmed by requests for assessments because they were so poorly staffed.[41] When conservation authorities tried to press for prosecutions for breaches of environmental legislation, they faced opposition from powerful sets of interest groups. For example, when the United Democratic Party were in power, the Minister of Tourism and Environment, became a developer in Placencia, where large areas of ecologically important mangroves were cut to make way for tourism developments.[42] Political opponents of such developments speculated that local business elites and external investors were able to seek protection from prosecution for environmental breaches due to their links with politically prominent figures in Belize, and their links to the ruling UDP.

The pace and direction of tourist development can also be shaped and driven by patterns of corruption. One of the major effects of corruption is that costs and benefits of any development are allocated in ways that benefit elites. Powerful politicians can be involved in shadow networks that include criminal organizations, which are then able to reshape the apparatus of the state to suit their own interests. Organized crime can enjoy protection from all levels of government, because it has invaded the structures of the state and taken advantage of its power and resources.[43] In this way global networks that often inextricably link legal and illegal businesses manage to incorporate developing states into the fabric of their organizations. The sector where these shadow networks are at their most pervasive and most effective in Central America is that of drug trafficking. The organizations and the networks involved in trafficking have not only incorporated the state apparatus, but infiltrated and reshaped the ways states are organized to protect and support their activities.

The Caribbean and Central America are two regions that have been targeted as trafficking routes by drug cartels. While Colombian drug cartels of the 1980s and early 1990s were highly visible and posed a direct threat to the state, by the late 1990s the nature of trafficking changed significantly, and its focus turned to

Mexico. The new breed of traffickers was more discreet and set up smaller scale organizations that are interested in co-opting key personnel within the political and judicial systems. Traffickers were also more likely to work through legitimate small businesses and contract out manufacturing and transport to specialist groups.[44] Within the Central American/Caribbean region, Belize has not been immune to the development of this illegal international trade in narcotics. The US Department of State identified Belize as a significant drug transit country. Since Belize lies between the producing countries of South America and the consumer countries of Europe and North America, its position marks it out as an ideal route for smugglers. In addition, the unique geography of the Caribbean, and especially Belize, means that chains of hundreds of islands provide points to drop off and pick up consignments of drugs.[45]

It is clear that on the one hand authorities in Belize were overwhelmed by the power of illicit networks, and on the other hand that elements in the formal state apparatus were engaged in illicit activities. For example, the US Bureau for International Narcotics and Law Enforcement Affairs stated that the ability of the government of Belize to combat trafficking was severely undermined by deeply entrenched corruption that reached into senior levels of government. In addition, it indicated that ministers in the government as well as police officers were involved in the drug trade in Belize.[46]

Such shadow networks are unlikely to be concerned about the need to conform to environmental legislation as they build resorts. As a result it is not uncommon for resorts to mean mangrove clearance, beach building and remodelling of the environment in ways that have a negative impact on wildlife. All these processes are made invisible to tourists; once a hotel opens its doors to guests the processes that have led to the construction of the hotel are hidden from view. Instead, tourists are much more likely to assume that they are looking out on a pristine landscape or an unspoilt beach. They do not see that turtle nesting sites have been disturbed, that the beach might be an artificial creation, that mangroves have been removed, people have been displaced, or that the money to build the hotel came from drug smuggling.

The links between tourism, conservation and criminal activity are obviously highly complex and stretch beyond buying souvenirs made from turtleshell. Clandestine linkages between politicians, developers, tour operators and criminal networks mean that the pristine tourist paradise we demand is a fabrication built on a completely redesigned and reshaped landscape. This then challenges the idea that tourism offers a pathway to sustainable development, a means of saving the environment while offering economic growth (especially to developing countries). Powerful interests at the local, national and international levels continue to promote tourism because they directly benefit from it; the benefits to poorer and more marginalized communities are harder to see. Government and business elites at the national level and international tour operators all benefit financially from the development of tourism in places like Belize. Once we understand the ways that tourist landscapes are produced, it becomes much harder to sustain the idea that tourism and tourists might be the saviours of wildlife.

CONCLUSION

Tourism is promoted as a potential solution to the problem of delivering economic development in an environmentally sustainable way, and this is a bewitchingly simple argument. Western tourists are particularly attracted by apparently pristine beaches, by back-to-nature experiences and by wildlife safaris. However, the assumption that this necessarily translates into the kinds of development that benefit wildlife is far too simplistic. The idea that tourism revenue is ploughed back into wildlife conservation to protect the animals and landscapes that tourists come to see is also far too simplistic. I do not contest the idea that in some places tourism is an important source of income for wildlife conservation. The massive and lucrative safari industry in Kenya and Tanzania are good examples of tourism activity that relies on wildlife being conserved, while gorilla-trekking permits in Rwanda and Uganda are an important source of government revenue.

However, the promotion of tourism=conservation by tour operators, government and conservation NGOs ignores the ways they are linked through crime. At one end of the scale, tourists can be both willing and unwitting contributors to environmental damage through their souvenir-buying habits. However, this is only part of the story. Holiday makers are mostly unaware of how their tourist paradises have been produced. They assume that the picture-perfect landscape or the silver Caribbean beach is a natural feature which has always existed. This is very far from the truth. Tourist playgrounds, like apparently pristine wildernesses, are manufactured environments, areas set aside by governments for wildlife, usually been cleared of people so that they conform to the global image of a wild landscape. Similarly, hotel construction in tropical areas can result in clearing ecologically important mangroves, or beach building which harms coral reefs. The complex inter-relationships between criminal networks involved in other forms of illicit activity such as drug smuggling has an effect on the environmental damage done by resort development. Money talks, and the links with powerful politicians mean that environmental regulations can be easily sidestepped.

CONCLUSION

BILLIONS OF DOLLARS ARE spent on saving species and habitats every year, but wildlife still appears to be in decline all over the world. This book started with a question: why is it that global conservation policies do not always achieve the desired effect? The answer lies in the failure to see conservation in its global context. This means that we focus on proximate causes of wildlife loss. Rather than tackling the ways demand for wildlife products is created and sustained by the wealthy world, conservation focuses on enforcement, coercion and regulation, which can have a negative and unjust impact on the world's poorest communities. Consequently, conservation produces exclusion, marginalization and even violence for local communities. This is counterproductive: it alienates the very communities that live with wildlife, the people we all rely on to make conservation work.

The wildlife trade is one of the most lucrative trades in the world, and the huge profits combined with a low risk of being caught means it has attracted the attentions of organized crime.

The case of cockling in Morecambe Bay, discussed in chapter one, demonstrates the ways in which conservation and organized crime are inter-linked. Demand from the rich world for cockles produced a gold-rush mentality on Morecambe sands; this attracted the attention of organized crime which met the demand via the provision of cheap illegal labour and illegal cockle fishing. This allowed crime networks to capture maximum profits from the trade. As we saw in the case of the worldwide trade in shell-fish, those involved in the illegal wildlife trade are not 'wildlife specialists'; they traffic valuable endangered species as one part of a wider portfolio of illicit activities. Wildlife can be moved along the same international trading routes used for drugs, people, car smuggling and people trafficking. Organizations like Interpol and the UK National Wildlife Crime Unit are well aware of these complex connections, but they are not the main target of most conservation organizations. Enforcing international laws on the wildlife trade has proved to be extremely difficult, partly because of the involvement of criminal networks that are mobile, flexible and able to avoid detection. But there is a second reason that enforcement has proved to be so elusive.

Conservationists still focus on poverty, claiming that poor people poach and trade wildlife to make a living. It is the rich world that demands ivory for ornaments, rhino horn for medicines or caviar to eat. If we blame local people for the problems facing wildlife, we are pointing the finger at the wrong people. The cases of gorillas in central Africa and lemurs in Madagascar indicate that the threats to wildlife do not necessarily come from local communities in the area, but by organized groups supported and driven by demand for minerals and gemstones in the wealthy world. Gorillas are threatened because of the history of conflict in the region, the current power dynamics and mineral extraction. The focus on local communities as 'the problem' makes it a localized issue, which can be fixed by focusing on the local area. This misses the wider dynamics of conflict and international trade in diamonds and strategic minerals.

It is important to tackle the consumption habits of the wealthy world rather than putting money into violent and coercive protection of national parks in the poorest parts of the world. Conservation organizations need to educate consumers so that they demand that any wildlife products they buy are produced under the highest ethical and environmental standards. Conservation agencies and NGOs find it difficult to show that their campaigns have a direct impact on buying habits. It is much easier to demonstrate impact if they train, equip and fund an anti-poaching patrol, fence a national park or train local tour guides. However, the collapse in demand for ivory in 1989 shows that consumer habits can be tackled. The 1989 ivory ban succeeded because demand had already shut down thanks to high-profile campaigns that linked poaching to the use of ivory for jewellery, billiard balls and piano keys. It is difficult for NGOs, especially, to turn their attention to their own supporters and point out their role in creating conservation problems; they risk alienating their donors, and supporters may not be willing to accept that their own lifestyles and demand for wildlife goods creates problems far distant from where they live. NGOs might well be acting on the best intentions and have the interests of wildlife at heart, but their failure to tackle the wider context means they perpetuate unhelpful stereotypes of poor people as the enemy of wildlife, and rich people as its saviour.

The case of the successful campaign for the ivory ban brings us to the second question tackled by this book: why do well-intentioned conservation initiatives pit local communities against wildlife? Part of the answer lies in the tendency to develop 'one-size-fits-all' conservation policies. These do not take account of variations in wildlife populations, which may be stable in one region, while threatened in others. They also ignore global inequalities in wealth and power, so that conservation feeds into the continuing dominance of the wealthy world. Elephant populations have recovered in Southern Africa, but it has been difficult to gain any support for alternative ways of thinking about elephant conservation. Southern African states argue that they are being punished by the ban; their substantial successes in elephant conservation are ignored, while the ban rewards East African countries for their failures. The ways that the Western public sentimentalize elephants means that the realities of living with them are made invisible. This is evident in the failure to understand why rural communities in Africa are hostile towards elephants that raid their crops or why national governments argue that they should be compensated for the loss of revenue because they are prevented from selling off stockpiles of ivory. If conservation fails to tackle the ways elephant management intersects with rural poverty, human–wildlife conflict and the constraints on national budgets then the elephant will be doomed in the longer term. The ivory trade ban is hailed as a conservation success, but the negative impact on people across Africa is simply ignored.

Conservation initiatives alienate local communities via their approach to wildlife management. Conservation rules have the capacity to determine and set definitions of what is criminal behaviour and what is not. At the global level, the international legal framework that governs conservation has the power to define legitimate and illegitimate use of wildlife. But decisions about what is legal, what is criminal, what is legitimate and what is illegitimate are not neutral scientific decisions. Instead these decisions are produced by political dynamics and a series of moral judgements about the acceptable limits to using wildlife. Conservation NGOs present themselves as trusted organizations that are committed to saving wildlife; they are 'above politics' and working in the true interest of species. They are uniquely placed to produce knowledge about wildlife and the impact of criminality, and they are highly skilled at campaigning and communicating their standpoints. However, conservation NGOs are political organizations with political agendas to promote.

The current approach to conservation produces exclusion, marginalization, and even violence. There seems to be an increasing turn back towards aggressive styles of 'fortress conservation', and an assumption that community conservation has failed. Concepts such as biodiversity and wilderness underpin current approaches to conservation, and have been exported all over the world via the creation of national parks and by international conservation NGOs. The problem is that wilderness is a Western fixation, and wilderness areas are manufactured for wildlife. Human communities have been removed, often with considerable violence.

The social injustices inherent in attempts to preserve species and ecosystems are at their sharpest in anti-poaching operations, which rely on aggressive protection and strict separation of human communities and wildlife. The definition of hunting as a crime against nature and future generations, coupled with the fact that international tourism has made some species and habitats financially valuable, provides a powerful rationale for the use of coercive and violent methods of protecting wildlife. The inherent problems in shoot-to-kill approaches are compounded by the arrival of private armies used to enforce protected areas for governments, NGOs and private philanthropists hoping to 'save' wildlife. In effect private companies are being given the right to take human life in defence of animals. We need to be clear about the concrete outcomes for people around the world when we demand that wildlife is protected. Sometimes it means that people will be killed only because they are *suspected* of wrongdoing. Consequently, conservation implies injustice for people all around the world. It is no surprise that they resist and challenge it.

Critics focus on the use of violence and on eviction, but, all around the world, people are being silently displaced in more insidious and invisible ways. They are excluded by laws banning grazing, agriculture, collection of plants for food or medicine, grazing and subsistence hunting in or near conservation areas. Everyday activities have been redefined as criminal acts, and local communities are branded as trespassers in protected areas. Such exclusions mean that poorer communities who rely on wild plants and animals to make ends meet, especially during times of crisis, are unable to feed themselves.

Tourism is promoted as the answer by conservation NGOs, the World Bank, tour operators and donors. The creation of opportunities in the tourism industry is offered as compensation for the loss of access to land and resources. But the idea that tourism will fill the gap and produce sustainable economic development is hopelessly naïve. For example, many elephants live in areas that will not attract international tourism; according to WWF 80 per cent of elephant range is outside national parks,[1] they share areas with human communities and in those areas the elephants are at a low density; but international tourists want guaranteed sightings of great herds of elephants living in a fantasy landscape that conforms to the image of an Edenic and (crucially) people-free wilderness.

Tourists themselves also contribute to the wildlife trade and wider forms of environmental change that damage wildlife. Western tourists want to see pristine beaches, engage in back-to-nature experiences and enjoy wildlife safaris in national parks, but the kind of development this involves does not necessarily benefit wildlife. Beaches are built, mangroves stripped out, waterholes drilled, people displaced and forests cleared to make way for tourist developments. The ways in which the environment has been changed and redesigned to conform to tourist expectations are largely invisible to visitors who only stay for a week or two. Tourism is another example of how richer communities drive these forms of environmental change, but poorer communities are blamed for unsustainably harvesting wood, lobsters and seashells. Our everyday lives are linked to conservation successes and failures across the world. Small changes in individual behaviours, choices about what products we buy and what kinds of conservation we support can make a big difference to wildlife.

To conclude, the wildlife trade has a major effect on wildlife populations around the world. However, blaming poverty is misplaced; without demand from richer communities around the world, poorer communities would not engage in poaching, smuggling and trading. The richer world is accustomed to consuming wildlife as accessories, as food, as medicine, as clothing and even as tourist attractions. The paradox is that conservation is itself implicated in developing and sustaining this situation. Creating new national parks or engaging local communities in conservation schemes inevitably means the development of new rules and regulations. These new rules effectively criminalize the everyday practices of local communities. Local people are redefined as illegal squatters, intruders and poachers in the drive to manufacture people-free wildernesses for wildlife. Local communities are promised that they will benefit from conservation: there will be new jobs in the tourism sector, they will be partners in the schemes and it will protect globally important species. But the reality is that the communities are rarely equal partners and very few jobs are created for local communities, and these tend to be poorly paid and seasonal. The failure to confront and tackle these very real issues on the ground means conservation will continue to get it wrong.

NOTES

INTRODUCTION

1. Baillie, J. E. M., C. Hilton-Taylor and S. N. Stuart (2004) *2004 IUCN Red List of Threatened Species: A Global Species Assessment* (Gland, Switzerland: IUCN). Estimates of the total number of species range from three million to a hundred million species on earth. The Convention on Biological Diversity operates on the assumption that there are approximately thirteen million species. Full text of the Convention on Biological Diversity is available at http://www.cbd.int (accessed 19.01.10).
2. Ehrlich, P. R. and E. O. Wilson (1991) 'Biodiversity studies: science and policy', *Science*, 253: 758–62.
3. To confuse matters further, approximately 15–20,000 new species are being described annually, and numerous species have been split into two or more new ones as our knowledge advances. Pimm et al. suggest that a high rate of loss is greater than one species lost per million species per year; see Pimm, S. L. and P. Raven (2000) 'Extinction by numbers', *Nature*, 403: 843–45; and Pimm, S. L. et al. (1995) 'The future of biodiversity', *Science*, 269: 347–50.
4. Myers, N. (2003) 'Viewpoint: Biodiversity Hotspots Revisited', *Bioscience*, 53(10): 916–17; and Myers, N. (1988) 'Threatened biotas: 'hot spots' in tropical forests', *The Environmentalist*, 8(3):187–208; Myers, N. (1990) 'The biodiversity challenge: expanded hot-spots analysis', *The Environmentalist*, 10(4):243–56.
5. http://www.wwf.org.uk/what_we_do/about_us/faqs/ (accessed 15.02.10).
6. http://wwf.worldwildlife.org/site/PageServer?pagename=can_gen_FAQ (accessed 15.02.10).
7. Chapin, M. (2004) 'A challenge to conservationists', *World Watch Magazine*, 17(6): 22; Scholfield, K. and D. Brockington (forthcoming 2010), 'Actually

225

existing conservation: understanding conservation NGOs in sub-Saharan Africa', *Antipode: A Journal of Radical Geography*.

8. http://www.nature.org/joinanddonate/corporatepartnerships/about/ (accessed 15.02.10).

9. http://www.savethetigerfund.org/AM/Template.cfm?Section=Home1 (accessed 12.02.10). Other figures indicate that WWF is the biggest spender in Africa at just over US$42 million per year, followed by Conservation International at US$22 million, Wildlife Conservation Society with US$17.3 million and African Wildlife Foundation at US$14.6 million. Scholfield and Brockington (forthcoming 2010) 'Actually existing conservation'; see also Borgerhoff-Mulder, M. and P. Coppolillo (2005) *Conservation: Linking Ecology, Economics and Culture* (Princeton, NJ: Princeton University Press) for further discussion of the changing profile of conservation.

10. http://www.rhinos-irf.org, International Rhino Foundation (accessed 08.04.09).

11. http://www.rspb.org.uk/ourwork/conservation/projects/eastscotlandeagles/index.asp (accessed 12.02.10); 'Record year for sea eagle pairs', (24.09.09) at http://news.bbc.co.uk/1/hi/scotland/highlands_and_islands/8270703.stm (accessed 12.02.10).

12. http://www.durrell.org/Animals/Birds/Mauritius-Pink-Pigeon/ (accessed 12.02.10); see also Butchart, S.H.M., A.J. Stattersfield and N.J. Collar (2006), 'How many bird extinctions have we prevented?', *Oryx*, 40(3): 266–78.

13. http://www.rhinos-irf.org, International Rhino Foundation (accessed 08.04.09).

14. http://www.panda.org/who_we_are/wwf_offices/cameroon/index.cfm?uProjectID=9F0725 (accessed 28.01.10); the ivory and rhino horn trades are discussed more fully in chapter 4.

15. http://www.iucnredlist.org/details/15955 (accessed 11.04.09).

16. Brockington, D. and J. Igoe (2006) 'Eviction for conservation: a global overview', *Conservation and Society*, 4(3): 443.

17. http://www.marineeducationtrust.org/petition/protect-chagos (accessed 22.02.10).

18. F. Pearce, 'The Chagos Archipelago, where conservation meets colonialism', *Guardian*, 18.02.10 at http://www.guardian.co.uk/environment/2010/feb/18/chagos-nature-reserve-greenwash. See also Sand, P. H. (2009) *The United States and Britain in Diego Garcia: The Future of a Controversial Base* (London: Palgrave/ Macmillan); and Vine, D.S. (2009) *Island of Shame: The Secret History of the US Military Base on Diego Garcia* (Princeton, NJ: Princeton University Press).

19. http://www.riotintomadagascar.com/development/ecotourism/index.html (accessed 10.11.07).

20. http://www.riotintomadagascar.com/ (accessed 10.11.07).

21. 'World Bank approves US$129 million for integrated growth poles in Madagascar' (12.07.05) at http://web.worldbank.org/WBSITE/EXTERNAL/COUNTRIES/AFRICAEXT/0,,contentMDK:20579808~menuPK:258658~pagePK:2865106~piPK:2865128~theSitePK:258644,00.html; interview with Serge Lachapelle, Director, Qit-Fer Minerals Madagascar (QMM), Antananarivo, 25.03.04.

22. http://www.riotintomadagascar.com/overview/impact/index.html (accessed 10.11.07).

23. http://www.foe.co.uk/resource/press_releases/rio_tintos_madagascar_mini_22102007.html (accessed 10.11.07).

24. http://www.traffic.org/trade (accessed 05.11.09).

25. For further discussion see Sanderson, S.E. and K.H. Redford (2003) 'Contested relationships between biodiversity conservation and poverty

alleviation?', *Oryx*, 37: 1–2; and Adams, W.M. et al. (2004) 'Biodiversity conservation and the eradication of poverty?', *Science*, 306: 1146–9.

26. Brockington and Igoe (2006) 'Eviction for conservation', pp. 424–70.

CHAPTER 1: THE INTERNATIONAL WILDLIFE TRADE

1. http://www.state.gov/g/oes/rls/prsrl/2007/81005.htm (accessed 28.04.08); and see Broad, S., T. Mulliken and D. Roe (2003) 'The Nature and Extent of Legal and Illegal Trade in Wildlife' in S. Oldfield (ed.), *The Trade in Wildlife: Regulation for Conservation* (London: Earthscan), pp. 3–22.

2. http://www.traffic.org/overview/ (accessed 19.01.10).

3. http://www.traffic.org/trade/ (accessed 28.04.08).

4. TRAFFIC (2008) *What's Driving the Wildlife Trade? A Review of Expert Opinion on Economic and Social Drivers of the Wildlife Trade and Trade Control Efforts in Cambodia, Indonesia, Lao PDR and Vietnam* (Cambridge: TRAFFIC International/ Washington DC: World Bank).

5. '*Crisis of Marine Plunder in Africa*', 02.10.07, Report by the Institute of Security Studies in South Africa; available at http://www.iss.co.za/index.php?link_ id=4056&slink_id=5023&link_type=12&slink_type=12&tmpl_id=3 (accessed 15.04.09).

6. Engler, M. and R. Parry-Jones (2007) *Opportunity or Threat: The Role of the European Union in Global Wildlife Trade* (Brussels: TRAFFIC-Europe), p. 33.

7. Engler and Parry-Jones (2007) *Opportunity or Threat*, pp. 24–5; and Knapp, A., C. Kitschke and S. von Meibom (eds) (2006) *Proceedings of the International Sturgeon Enforcement Workshop to Combat Illegal Trade in Caviar* (Brussels: TRAFFIC-Europe).

8. 'London raids expose Mafia caviar racket', *The Observer*, 09.11.03; 'Poachers' hunger for black gold may take caviar off the menu', *The Independent*, 20.06.01.

9. South African Government application for an Appendix III listing for CoP 14 of CITES, available at http://www.cites.org/common/cop/14/inf/E14i-58.pdf (accessed 14.04.09).

10. 'Rare shellfish bartered for drugs', *The Observer* (UK) 23.09.07; and Bürgener, M. (2005), 'South African Abalone: A CITES Appendix III Candidate?' *TRAFFIC Bulletin*, 20 (2): 48–9.

11. South African Government application for an Appendix III listing for CoP 14 of CITES, available at http://www.cites.org/common/cop/14/inf/E14i-58.pdf (accessed 14.04.09); also see Bürgener, M. (2005), 'South African Abalone: A CITES Appendix III Candidate?' *TRAFFIC Bulletin*, 20(2): 48–9.

12. Steinberg, J. (2005) *The Illicit Abalone Trade in South Africa* (Johannesburg: Institute for Security Studies).

13. 'Rare shellfish bartered for drugs', *The Observer* (UK) 23.09.07; see also Kapp, C. (2008) 'Crystal meth boom adds to South Africa's health challenges', *The Lancet*, 371 (9608): 193–4.

14. Rangarajan, M. 'Can the tiger save India?', *The Deccan Chronicle*, 08.01.10.

15. http://www.traffic.org/tigers (accessed 11.04.09) also see http://www.panda. org/what_we_do/endangered_species/endangered_species_list/tigers/tiger_sol utions/challenges/ (accessed 14.04.09).

16. http://www.iucnredlist.org/details/15955 (accessed 11.04.09).

17. Nowell, K. and L. Xu (2007) *Taming the Tiger Trade: China's Markets for Wild and Captive Tiger Products since the 1993 Domestic Trade Ban* (Hong Kong: TRAFFIC East Asia).

18. Nowell, K. (2007) *Asian Big Cat Conservation and Trade Control in Selected Range States: Evaluating Implementation and Effectiveness of CITES Recommendations* (Cambridge, UK: TRAFFIC International).
19. EIA (2008) *Skin Deep* (London: EIA), p. 3.
20. http://www.eia-international.org/campaigns/species/tigers/ (accessed 09.04.09).
21. See Williamson, D. F. and L. A. Henry (2008) *Paper Tigers? The Role of the US Captive Tiger Population in the Trade in Tiger Parts* (Washington DC: TRAFFIC US) for further discussion.
22. http://www.iucnredlist.org/details/15955 (accessed 11.04.09)
23. http://www.savethetigerfund.org (accessed 09.04.09).
24. http://www.hmrc.gov.uk/ (accessed 08.08.08).
25. Interview with Project Officer, Regional Law Enforcement Assistance, ASEAN-WEN Support Programme, TRAFFIC South East Asia, Bangkok, 06.03.08.
26. http://www.nwcu.police.uk/ (accessed 04.11.08); interview with Inspector Brian Stuart, National Wildlife Crime Unit, North Berwick, 22.05.09.
27. http://www.defra.gov.uk/paw/default.htm (accessed 08.08.08). http://www.ukcites.gov.uk/enforce/default.htm (accessed 08.08.08).
28. http://www.operationcharm.org/news/20070709.jsp (accessed 19.01.10).
29. http://www.operationcharm.org/ (accessed 08.08.08); and see http://www.met.police.uk/wildlife/operation_charm.htm(accessed 08.08.08).
30. http://www.operationcharm.org/celebrities/ (accessed 19.01.10).
31. http://www.interpol.int/Public/EnvironmentalCrime/Wildlife/Default.asp (accessed 08.08.08).
32. IFAW (2008) *Killing With Keystrokes: An Investigation of the Illegal Wildlife Trade on the World Wide Web* (Brussels: IFAW), p. 2.
33. IFAW (2008) *Killing With Keystrokes*.
34. 'Evil gangmasters who rule the cockle slave trade by fear', *The Guardian*, 08.02.04.
35. Personal comment, police officer in Lancaster, 08.09.08; also see interview with Geraldine Smith, MP for Morecambe and Lunesdale, Morecambe, 25.09.08.
36. 'Cockler gangmaster jailed for 14 years', *The Guardian*, 28.03.06.
37. Interview with Geraldine Smith, 25.09.08
38. 'I got calls from people in the sea. The tide was coming up', *The Guardian*, 25.03.08
39. Interview with Geraldine Smith, 25.09.08; personal comment, Morecambe bay fisherman, 22.09.08; and personal comment, police officer in Lancaster, 08.09.08.
40. 'Tell the family to pray for me . . . I am dying', *The Guardian* 25.03.08.
41. Interview with Geraldine Smith, 25.09.08.

CHAPTER 2: GLOBAL ACTION, LOCAL COSTS

1. Roe, D. (2008) *Trading Nature: A Report with Case Studies, on the Contribution of Wildlife Trade Management to Sustainable Livelihoods and the Millennium Development Goals* (Cambridge, UK: TRAFFIC International and WWF International), pp. 68–9.
2. http://www.cites.org (accessed 17.12.08); Dickson, B. (2003) 'What is the goal of regulating the wildlife trade? Is regulation a good way to achieve this goal?' in S. Oldfield (ed.) *The Trade in Wildlife: Regulation for Conservation* (London: Earthscan) pp. 23–32; Cooney, R. and P. Jepson, P. (2006) 'The international wild bird trade: What's wrong with blanket bans?', *Oryx* 40(1): 18–23; and

Roe, D. (2006) 'Blanket bans: conservation or imperialism? A response to Cooney & Jepson' *Oryx*, 40(1): 27–8.

3. http://www.cites.org (accessed 17.12.08).

4. For further discussion see Hutton, J. and B. Dickson (2000) *Endangered Species, Threatened Convention: The Past Present and Future of CITES* (London: Earthscan).

5. Litfin, K. (1994) *Ozone Discourses: Science and Politics in Global Environmental Co-Operation* (New York: Columbia University Press), p. 51; see also Demerrit, D. (2006) 'Science studies, climate change and the prospects for a constructivist critique', *Economy and Society*, 35(3): 453–79.

6. For further discussion of CITES see Oldfield, S. (ed.) (2003) *The Trade in Wildlife: Regulation for Conservation* (London: Earthscan) and Hutton, J. and B. Dickson (2000) *Endangered Species, Threatened Convention: The Past Present and Future of CITES* (London: Earthscan).

7. See Hutton, J. and B. Dickson (2000) *Endangered Species, Threatened Convention: The Past Present and Future of CITES* (London: Earthscan) for further discussion.

8. Dickson, B. (2003) 'What is the Goal of Regulating the Wildlife Trade? Is Regulation a Good Way to Achieve This Goal?' in S. Oldfield, (ed) *The Trade in Wildlife*, pp. 23–32; and Hutton, J. and B. Dickson (2000) *Endangered Species*.

9. See http://www.fws.gov/laws/lawsdigest/LACEY.HTML (accessed 01.02.10); http://www.fws.gov/Endangered/ (accessed 01.02.10).

10. Interview with conservation professional in Madagascar who preferred to remain anonymous (March 2004).

11. Wilson, E. O. (ed.) (1988) *Biodiversity* (Washington DC: National Academy of Sciences/Smithsonian Institution) and Wilson, E. O. (1992) *Diversity of Life* (Cambridge, MA: Harvard University Press).

12. Adams, W. M. (2004) *Against Extinction: The Story of Conservation* (London: Earthscan), pp. 5–27.

13. See http://www.eowilson.org (accessed 04.11.08).

14. Full text of the Convention on Biological Diversity is available at http://www.cbd.int (accessed 19.01.10).

15. Demeritt (2006) 'Science studies, climate change; Litfin (1994) *Ozone Discourses*.

16. Adams (2004) *Against Extinction*, pp. 105–8.

17. http://www.krugerpark.co.za/Kruger_National_Park_Lodging_&_Camping_Guide-travel/skukuza-camp.html (accessed 06.08.08); and Carruthers, J. (1995) *The Kruger National Park: A Social and Political History* (Pietermaritzburg: University of Natal Press); Fabricius, C. and de Wet, C. (2002) 'The influence of forced removals and land restitution on conservation in South Africa', in D. Chatty and M. Colchester (eds) *Conservation and Mobile Indigenous Peoples. Displacement, Forced Settlement, and Sustainable Development* (New York: Berghahn Books).

18. Jacoby, K. (2003) *Crimes Against Nature: Squatters, Poachers, Thieves, and the Hidden History of American Conservation* (Berkeley, CA: University of California Press); see also Dowie, M. (2009) *Conservation Refugees: The Hundred-Year Conflict between Global Conservation and Native Peoples* (Cambridge, MA: MIT Press), pp. 1–15 and R. P. Neumann (1998) *Imposing Wilderness: Struggles over Livelihood and Nature Preservation in Africa* (Berkeley, CA: University of California Press).

19. Jacoby, K. (2003) *Crimes Against Nature: Squatters, Poachers, Thieves, and the Hidden History of American Conservation* (Berkeley, CA: University of California Press) p. 2.

229

20. Dowie, M. (2009) *Conservation Refugees: The Hundred-Year Conflict between Global Conservation and Native Peoples* (Cambridge: MIT Press), pp. 1–15; and see Adams, J. S. and McShane, T. O. (1992) *The Myth of Wild Africa: Conservation Without Illusion* (Berkeley, CA: University of California Press); and Garland, E. (2008) 'The Elephant in the Room: Confronting the Colonial Character of Wildlife Conservation in Africa' *African Studies Review*, 51(3): 51–74

21. Brockington, D., R. Duffy and J. Igoe (2008) *Nature Unbound: Conservation, Capitalism and the Future of Protected Areas* (London: Earthscan), p. 19.

22. Brockington et al. (2008) *Nature Unbound*, p. 24.

23. Jacoby, K. (2003) *Crimes Against Nature: Squatters, Poachers, Thieves, and the Hidden History of American Conservation* (Berkeley, CA: University of California Press) p. 3; see also Forsyth, T. and A. Walker (2008) *Forest Guardians, Forest Destroyers: The Politics of Environmental Knowledge in Northern Thailand* (Seattle, WA: University of Washington Press).

24. Dowie, M. (2009) *Conservation Refugees: The Hundred-Year Conflict between Global Conservation and Native Peoples* (Cambridge: MIT Press), pp. 1–15.

25. http://www.africanparks-conservation.com/apffoundation/index.php (accessed 16.11.09); http://www.africanparks-conservation.com/apffoundation/index.php (accessed 16.11.09); Pearce, F. (2005b) 'Laird of Africa'. *New Scientist*, 13 August 2005: 48–50.

26. http://www.africanparks-conservation.com/apffoundation/index.php (accessed 16.11.09); http://www.refugeesinternational.org/content/photo/detail/4727/ (accessed 16.11.09); http://www.refugeesinternational.org/content/article/detail/ 56http://conservationrefugees.org/NechSar (accessed 16.11.07).

27. http://www.conservationrefugees.org/ (accessed 06.08.08).

28. Brockington, D. and J. Igoe (2006) 'Eviction for conservation: a global overview', *Conservation and Society*, 4(3): 424–70.

29. Robbins, P., K. McSweeney, A. K. Chhangani and J. L. Rice (2009) 'Conservation as it is: Illicit resource use in a wildlife reserve in India', *Human Ecology*, 37(5): 559–75.

30. Brockington and Igoe (2006) 'Eviction for conservation': Borgerhoff-Mulder, M. and P. Coppolillo (2005) *Conservation: Linking Ecology, Economics, and Culture* (Princeton, NJ: Princeton University Press).

31. Mbaiwa, J. (2005) 'Wildlife resource utilisation at Moremi Game Reserve and Khwai Community Area in the Okavango Delta, Botswana', *Journal of Environmental Management*, 77: 144–56; and Mbaiwa, J.E. (2005) 'Enclave Tourism and its Socio-Economic Impacts in the Okavango Delta, Botswana' *Tourism Management*, 26: 157–72; and see http://www.survivalinternational. org/tribes/bushmen (accessed 16.03.10).

32. Adams, W. M. (2004) *Against Extinction: The Story of Conservation* (London: Earthscan), p. 24.

33. Duffy, R. (2006) 'Global governance and environmental management: The politics of transfrontier conservation areas in Southern Africa', *Political Geography*, 25(1), p. 102; Wolmer, W. (2006) *From Wilderness to Farm Invasions: Conservation and Development in Zimbabwe's Southeast Lowveld* (Oxford: James Currey); and see Ali, S. (2007) *Peace Parks Conservation and Conflict Resolution* (Cambridge, MA: The MIT Press); Büscher, B. (2010). Seeking Telos in the 'Transfrontier': Neoliberalism and the Transcending of Community Conservation in Southern Africa. *Environment and Planning A* 42 (3): 644–660; Büscher, B. (2010) Anti-Politics as Political Strategy: Neoliberalism and Transfrontier Conservation in Southern Africa. *Development and Change* 41(1): 29–51 for further discussion.

34. West, P. and J. G. Carrier (2004) 'Ecotourism and authenticity. Getting away from it all?', *Current Anthropology*, 45(4): 483–98.

35. Conor Foley, 'Not cool', *The Guardian*, 12.06.08; http://www.coolearth.org/394/category/save-an-acre-182.html (accessed 27.01.10); 'Tory backer to advise labour on energy and deforestation', *The Guardian*, 08.11.07; 'Johan Eliasch, Gordon Brown consultant, fined for illegal Amazon logging', *The Daily Telegraph*, 07.06.08.

36. For an excellent discussion of the role of celebrities in conservation see Brockington, D. (2009) *Celebrity and the Environment: Fame, Wealth and Power in Conservation* (London: Zed Books).

37. For further discussion of the changing role of NGOs in a new world order see Hulme, D. and M. Edwards (1997) *NGOs, States and Donors: Too Close for Comfort?* (London: Macmillan).

38. Chapin, M. (2004) 'A challenge to conservationists', *World Watch Magazine*, 17(6): 17–31, p. 22; Scholfield, K. and D. Brockington (forthcoming 2010) 'Actually existing conservation: Understanding conservation NGOs in sub-Saharan Africa', *Antipode: A Journal of Radical Geography*.

39. Chapin (2004) 'A challenge to conservationists', p. 23; Halpern, B. S. et al. (2006), 'Gaps and mismatches between global conservation priorities and spending', *Conservation Biology* 20(1): 56–64.

40. Scholfield and Brockington (forthcoming 2010), 'Actually existing conservation'.

41. http://www.conservation.org/campaigns/starbucks/Pages/default.aspx (accessed 19.01.08); http://www.conservation.org/discover/partnership/corporate/Pages/mcdonalds.aspx (accessed 19.01.08).

42. Pawliczek, M. and H. Mehta (2008) 'Ecotourism in Madagascar: How a sleeping beauty is finally awakening' in A. Spenceley (ed.), *Responsible Tourism: Critical Issues for Conservation and Development* (London: Earthscan), pp. 41–68.

43. Brockington, D. (2009) *Celebrity and the Environment: Fame, Wealth and Power in Conservation* (London: Zed Books), p. 90; Corson, C. (forthcoming 2010) 'Shifting environmental governance in a neoliberal world: The politics of US(aid) for conservation', *Antipode: A Journal of Radical Geography*.

44. See Vira, B. and W.M. Adams (2009) 'Ecosystem services and conservation strategy: beware the silver bullet', *Conservation Letters*, 2: 158–62; Redford, K. H. and W. A. Adams (2009) 'Payment for ecosystem services and the challenge of saving nature', *Conservation Biology*, 23(4): 785–87.

45. Goldman, M., (2001) 'Constructing an environmental state: eco-governmentality and other practices of a 'Green' World Bank', *Social Problems*, 48: 502–6; also see Goldman, M. (2005) *Imperial Nature: The World Bank and Struggles for Social Justice in the Age of Globalization* (New Haven, CT: Yale University Press).

46. Goldman (2001) 'Constructing an environmental state' and Goldman (2005) *Imperial Nature*; Agrawal, A. (2005) 'Environmentality: community, intimate government and the making of environmental subjects in Kumaon, India', *Current Anthropology*, 46(2): 161–90

47. Li, T. (2007) *The Will to Improve: Governmentality, Development, and the Practice of Politics* (Durham: Duke University Press), p. 164; also see Barrow, E. and Fabricius, C. (2002) 'Do rural people really benefit from protected areas – rhetoric or reality?', *Parks* 12: 67–79; and Bolaane, M. (2004) 'The impact of game reserve policy on the River BaSarwa/Bushmen of Botswana' *Social Policy and Administration* 38(4): 399–417.

48. Li (2007) *The Will to Improve*, p. 196.

49. Brockington, D., R. Duffy and J. Igoe (2008) *Nature Unbound: Conservation, Capitalism and the Future of Protected Areas* (London: Earthscan), p. 149.
50. Brockington et al. (2008), *Nature Unbound*, p. 149; see also Tsing, A. L. (2004) *Friction: An Ethnography of Global Connection* (Princeton, NJ: Princeton University Press).
51. For an excellent discussion of the history of Madagascar see Randrianja, S. and S. Ellis (2009) *Madagascar: A Short History* (London: Hurst and Company).
52. For further discussion of threats to biodiversity in Madagascar see http://www.bbc.co.uk, *'Madagascar Biodiversity Threatened'* (16.01.02; accessed 08.02.02); and 'Madagascar's jewels of nature under threat', *Financial Times*, 15.05.01.
53. Kull, C. A. (1996) 'The evolution of conservation efforts in Madagascar', *International Environmental Affairs* 8(1): 50–86; Kull, C. A. (2002) 'Madagascar's burning issue: the persistent conflict over fire', *Environment* 44(3): 8–19; Marcus, R. R. and C. A. Kull (1999) 'Setting the stage: The politics of Madagascar's environmental efforts', *African Studies Quarterly*, 3(2): 1–7.
54. *'Madagascar to Triple Areas Under Protection'* (16.09.03) http://www.conservation.org/xp/news/press_releases/020403.xml (accessed 03/11/03); see also http://www.conservation.org/xp/CIWEB/regions/africa/madagascar/ (accessed 17.11.04).
55. *An Overview of CEPF's Portfolio in the Madagascar and Indian Ocean Islands Biodiversity Hotspot: Madagascar*. http://www.cepf.net/ImageCache/cepf/content/pdfs/cepf_2emadagascar_2eoverview_5f3_2e05_2epdf/v1/cepf.madagascar.overview_5f3.05.pdf (accessed 04.07.07).
56. Horning, N. R. (2008) 'Strong support for weak performance: Donor competition in Madagascar', *African Affairs*, 107(428): 405–31.
57. Hutton, J., W. M. Adams and J. C. Murombedzi (2005) 'Back to the barriers? Changing narratives in biodiversity conservation', *Forum for Development Studies*, 32(2): 341–70; Neumann, R. P. (2004) 'Moral and discursive geographies in the war for biodiversity in Africa', *Political Geography*, 23: 813–37.

CHAPTER 3: WILDLIFE WARS: POACHING AND ANTI-POACHING

1. http://www.ifaw.org/ifaw/general/default.aspx?oid=32904 (accessed 25.06.08).
2. http://www.eia-international.org/old-reports/Tigers/Reports/StatTig/poach_trade01.html (accessed 25.06.08).
3. Neumann, R. P. (2004) 'Moral and discursive geographies in the war for biodiversity in Africa', *Political Geography*, 23: 813–37, p. 830.
4. MacKenzie, J. (1988) *Empire of Nature: Hunting, Conservation and British Imperialism* (Manchester: Manchester University Press).
5. Adams, W. M. (2004) *Against Extinction: The Story of Conservation* (London: Earthscan).
6. Adams (2004) *Against Extinction*, pp. 33–41; Neumann (2004) 'Moral and discursive geographies', p. 830.
7. Neumann, R. P. (2004) 'Moral and discursive geographies', p. 825.
8. See MacKenzie (1988) *Empire of Nature*.
9. 'Gangs moving into environment crime', *The Daily Telegraph*, 06.06.07.
10. Ng, J. and Nemora (2008) *The Tiger Trade Revisited in Sumatra, Indonesia* (Cambridge: TRAFFIC); 'Tiger poaching still at danger level', 29.03.2000. http://news.bbc.co.uk/1/hi/sci/tech/693802.stm (accessed 07.05.08)

11. 'Tiger poaching ring busted by Indian police', 06.12.08 at http://news.national-geographic.com/news/2007/12/071206-AP-india-tiger.html (accessed 07.05.08).
12. 'Big cats under threat: tiger tiger burning bright', *The Independent*, 15.02.06.
13. 'Health fears grow as mountains of meat are smuggled into the UK', *The Observer* 10.07.07; interview with TRAFFIC representative, who preferred to remain anonymous, Cambridge 13.05.09; and interview with Inspector Brian Stuart, National Wildlife Crime Unit, North Berwick, 22.05.09.
14. 'Hunters leave bonobo on the brink of extinction', *The Guardian*, 09.12.04.
15. 'Fears for East African bushmeat slaughter', *The Guardian*, 23.01.08; see also Bowen-Jones, E. (2003) 'Bushmeat: Traditional Regulation or Adaptation to Market Forces' in S. Oldfield (ed.), *The Trade in Wildlife: Regulation for Conservation* (London: Earthscan), pp. 132–45.
16. http://www.bushmeat.org/ (accessed 05.05.08).
17. Bonner, R. (1993) *At the Hand of Man: Peril and Hope for Africa's Wildlife* (London: Simon and Schuster), p. 54.
18. 'SA Army's Ivory Trail Exposed', *The Independent*, 23.3.92; and Reeve, R. and S. Ellis (1995) 'An Insider's Account of the South African Security Forces Role in the Ivory Trade' *Journal of Contemporary African Studies*, 13: 222–43.
19. 'SA Army's ivory trail exposed', *The Independent*, 23.3.92.
20. 'Report takes WWF to "Tusk" ', *Mail and Guardian*, 19.1.96.
21. Reeve, R. and S. Ellis (1995) 'An insider's account of the South African Security Force's role in the ivory trade', *Journal of Contemporary African Studies*, 13: 222–43; Duffy, R. (1999) 'The role and limitations of state coercion: anti-poaching policies in Zimbabwe', *Journal of Contemporary African Studies*, 17: 97–121.
22. Neumann, R. P. (2004) 'Moral and discursive geographies', p. 828.
23. See Neumann, R. (2001) 'Disciplining Peasants in Tanzania: From State Violence to Self Surveillance in Wildlife Conservation', in N. L. Peluso and M. Watts (eds), *Violent Environments* (Ithaca, NY and London: Cornell University Press), pp. 305–27; see also Neumann R. P. (1998) *Imposing Wilderness: Struggles over Livelihood and Nature Preservation in Africa* (Berkeley: University of California Press); and Dzingirai, V. (2003) 'The new scramble for the African countryside' *Development and Change* 34(2): 243–63; and Murombedzi, J. (2001) 'Why wildlife conservation has not economically benefitted communities in Africa', in D. Hulme and M. Murphree (eds) *African Wildlife and Livelihoods* (Oxford: James Currey) pp. 208–26.
24. Hutton, J., W. M. Adams and J. C. Murombedzi (2005) 'Back to the barriers? Changing narratives in biodiversity conservation', *Forum for Development Studies*, 32 (2): 341–70.
25. Terborgh, J. (1999) *Requiem for Nature* (Washington DC: Island Press); Oates, J. F. (1999) *Myth and Reality in the Rainforest: How Conservation Strategies Are Failing West Africa* (Berkeley: University of California Press); and Brandon, K., Redford, K. H. and Sanderson, S. E. (1998) *Parks in Peril: People, Politics and Protected Areas* (Washington DC: The Nature Conservancy).
26. Leakey, R. (2001) *Wildlife Wars: My Battle to Save Kenya's Elephants* (London: Pan), p. 48.
27. Leakey (2001) *Wildlife Wars*, p. 38.
28. Leakey (2001) *Wildlife Wars*, p. 102.
29. Fossey, D. (1983) *Gorillas in the Mist* (Boston, MA: Houghton Mifflin).
30. de la Bedoyere, C. (2005) *No One Loved Gorillas More* (Bath: Palazzo Editions).
31. www.rwandatourism.com/parks.htm (accessed 12.05.08).

32. Hayes, H. T. P. (1989) *The Dark Romance of Dian Fossey* (London: Simon and Schuster).
33. Weber, B. and A. Veder (2001) *In the Kingdom of Gorillas: Fragile Species in a Dangerous Land* (London: Simon and Schuster).
34. 'Parks units in raging battle', *Herald*, 5.11.87; 'Two killed as anti poaching war hots up', *Herald*, 9.2.87; see also Duffy, R. (1999) 'The role and limitations of state coercion: anti-poaching policies in Zimbabwe' *Journal of Contemporary African Studies*, 17: 97–121.
35. Wildlife Direct Website http://wildlifedirect.org/about.php# (accessed 12.05.08); see also 'The power of blogging helping Masai Mara wildlife', *The Daily Telegraph*, 03.03.08.
36. 'Swiss bid to check rhino slaughter', *Herald*, 14.3.88; 'Eighth poacher shot dead', *Herald*, 16.5.88. The helicopter was originally given by the Goldfields Trust to the Duke of Edinburgh in his capacity of WWF-International's President, and then it was assigned to Zimbabwe.
37. Bonner, R. (1993) *At the Hand of Man*, p.77.
38. 'WWF to halt copter aid', *Financial Gazette*, 31.3.89.
39. Ferguson, J. (2006) *Global Shadows: Africa in the Neoliberal World* (Durham, NC: Duke University Press), p. 49.
40. Jacoby, K. (2003) *Crimes Against Nature Squatters, Poachers, Thieves, & the Hidden History of American Conservation* (Berkeley, CA: University of California Press), p. 2.

CHAPTER 4: RHINO HORN, IVORY AND THE TRADE BAN CONTROVERSY

1. For further discussion see Rivalan, P., V. Delmas, E. Angulo, et al. (2007) 'Can bans stimulate wildlife trade?' *Nature*, 447: 529–30.
2. http://www.rhinos-irf.org, International Rhino Foundation (accessed 08.04.09).
3. http://www.savetherhino.org/eTargetSRINM/site/848/default.aspx (accessed 08.04.09).
4. Bradley Martin, E. and C. Bradley Martin (1982) *Run Rhino Run* (London: Chatto and Windus), pp. 66–75.
5. Bradley and Bradley Martin (1982) *Run Rhino Run*, pp. 53–5.
6. Mills, A. (1997) *Rhinoceros Horn and Tiger Bone in China: an Investigation of Trade Since the 1993 Ban* (Cambridge, UK: TRAFFIC).
7. IUCN (1980) *The International Trade in Rhinoceros Products*, Report Prepared by E. B. Martin for IUCN and WWF (Gland, Switzerland: IUCN/WWF).
8. EIA (1994) *CITES: Enforcement Not Extinction* (London: EIA), p. 4; see also Leader-Williams, N. (2003) 'Regulation and Protection: Successes and Failure in Rhinoceros Conservation' in S. Oldfield, (ed.), *The Trade in Wildlife: Regulation for Conservation* (London: Earthscan), pp. 89–99.
9. IUCN (1992) *Analysis of Proposals to Amend the CITES Appendices*. Prepared by the IUCN Species Survival Commission Trade Specialist Group, TRAFFIC and WCMC for the 8th Meeting of the Conference of the Parties, Kyoto, Japan, 2–13 March 1992 (Gland, Switzerland: IUCN), pp. 75–6.
10. EIA, David Shepherd Foundation and Tusk Force (1993) *Taiwan Kills Rhinos With Your Money: Why You Should Boycott Goods Made in Taiwan* (London: EIA).
11. http://www.fws.gov/laws/lawsdigest/RHINO.htm (accessed 01.02.10).

12. IUCN (1980) *The International Trade in Rhinoceros Products*, p. 26.
13. Duffy, R. (2000) *Killing for Conservation: The Politics of Wildlife in Zimbabwe* (Oxford: James Currey/Indiana: Indiana University Press).
14. For further discussion see Leader-Williams, N, S. Milledge, K. Adcock et al. (2005) 'Trophy hunting of black rhino Diceros bicornis: proposals to ensure its future sustainability', *Journal of International Wildlife Law and Policy*, 8: 1–11.
15. http://www.savetherhino.org/etargetsrinm/site/740/default.aspx (accessed 09.04.09).
16. 'Marthinuis puts a freeze on rhino horn trading', *Cape Argus*, 09.06.08.
17. 'Horn trade pressures some rhinos', 06.06.07, http://news.bbc.co.uk/1/hi/sci/tech/6726569.stm (accessed 09.04.09).
18. http://www.sanparks.org/parks/kruger/elephants/about/faq.php (accessed 31.07.09).
19. MacKenzie, J. M. (1988) *Empire of Nature: Hunting Conservation and British Imperialism* (Manchester: Manchester University Press).
20. Adams, W. M. (2004) *Against Extinction: The Story of Conservation* (London: Earthscan), p. 30; MacKenzie (1988) *Empire of Nature*.
21. http://www.panda.org/who_we_are/wwf_offices/cameroon/index.cfm?uProjectID=9F0725 (accessed 28.01.10).
22. Bonner, R. (1993) *At the Hand of Man: Peril and Hope for Africa's Wildlife* (London: Simon and Schuster), pp. 53–4.
23. EIA (1989) *A System of Extinction: The African Elephant Disaster* (London: EIA).
24. Currey, D. and H. Moore (1994) *Living Proof: African Elephants. The Success of the CITES Appendix I Ban* (London: EIA), p. 11.
25. 'The Elephant culling debate', *The Times* (South Africa) 01.03.08 at http://www.thetimes.co.za/PrintEdition/Insight/Article.aspx?id=717901 (accessed 18.06.09).
26. Moore, L. (2010) 'Conservation heroes versus environmental villains: perceiving elephants in Caprivi, Namibia', *Human Ecology*, 38(1): 19–29.
27. Hutton, J., W. M. Adams and J C. Murombedzi (2005) 'Back to the barriers? Changing narratives in biodiversity conservation' *Forum for Development Studies*, 32 (2): 341–70; Mukamuri, B. B. J. M. Manjengwa & S. Anstey (eds) (2009) *Beyond Proprietorship: Murphree's Laws on Community-Based Natural Resource Management in Southern Africa* (Harare: Weaver Press/Ottawa: International Development Research Centre).
28. Bonner, R. (1993) *At the Hand of Man: Peril and Hope for Africa's Wildlife* (London: Simon and Schuster), pp. 148–51.
29. 'Kenya in gesture burns ivory tusks', *New York Times*, 19.07.89; see also Leakey, R. (2001) *Wildlife Wars: My Battle to Save Kenya's Elephants* (London: Pan).
30. Bonner (1993) *At the Hand of Man*, pp. 148–51.
31. Bonner (1993) *At the Hand of Man*, pp. 148–51.
32. Duffy, R. (2008) 'Ivory Trade: Enforceability, Effectiveness and Ethical concerns' in C. Wemmer and K. Christen (eds), *Elephants and Ethics: The Morality of Coexistence* (Baltimore, MD: Johns Hopkins University Press), pp. 451–67.
33. Ivory Trade Review Group (1989) *The Ivory Trade and the Future of the African Elephant*, Vol. I (Summary and Conclusions). Report prepared for the 7th CITES, Conference of the Parties, Lausanne October 1989 (Oxford: ITRG), pp. 1–4.

34. Eugene Lapointe, Secretary General of CITES speaking at the Symposium on the *Conservation and Sustainable Use of Natural Resources*, held in Tokyo, 1 October 1993, Global Guardian Trust.

35. Bonner (1993) *At the Hand of Man*, pp. 140–8.

36. Thornton, A. and D. Currey (1991) *To Save an Elephant* (London: Bantam Books). p. 254.

37. Bonner (1993) *At the Hand of Man*, p. 159; Milliken, T. (1994) 'Current Trade in, Demand for and Stockpiles of Ivory in Range and Consumer States'. Paper presented at *The African Elephant in the Context of CITES*. Conference held in Kasane, Botswana 19–23 September 1994 (UK Department of Environment), pp. 49–54.

38. Bonner (1993) *At the Hand of Man*.

39. Hulme, D. and M. Murphree (eds) (2001) *African Wildlife and Livelihoods. The Promise and Performance of Community Conservation* (Oxford: James Currey).

40. Interview with Department of National Parks and Wildlife Management official (anonymised), Harare April 1995.

41. See Mukamuri, B. B, J. M. Manjengwa and S. Anstey (eds) (2009) *Beyond Proprietorship: Murphree's Laws on Community-Based Natural Resource Management in Southern Africa* (Harare: Weaver Press/Ottawa: International Development Research Centre); and Western, D. and R. Wright (eds.) (1994) *Natural Connections: Perspectives on Community Based Conservation* (Washington D.C.: Island Press) for further discussion.

42. Interview with Rowan Martin, Assistant Director: Research, DNPWLM, 29.5.95, Harare.

43. 'Animal rights outrage over plan to cull South Africa's elephants', *The Times*, 26.02.08 http://www.timesonline.co.uk/tol/news/world/africa/article3431369.ece (accessed 19.06.09); figures on elephant numbers in South African national parks can also be found at http://www.sanparks.org/parks/kruger/elephants/about/faq.php (accessed 03.07.09).

44. See http://www.huntinafrica.com/packages/packages_fixed.asp?page=packages& package_type=2; and http://www.phasa.co.za/index.php?pid=3 (accessed 19.06.09).

45. 'Animal rights outrage', *The Times*, 26.02.08.

46. 'Controversial elephant ivory auction begins in southern Africa', *The Guardian*, 28.10.08.

47. Richard Leakey, Wildlife Direct Press release 05.11.08 'Dr Richard Leakey denounces ivory auctions: one off sales will open up illegal markets'. http://safaritalk.net/index.php?s=3a7426980729064e4fbbab3aa7d874ef&show topic=2687&pid=13721&st=20&#entry13721 (accessed 16.02.10).

48. For a full description of IUCN Categories of Protected Areas see http://www.iucn.org/about/union/commissions/wcpa/wcpa_work/wcpa_strategic/wcpa_science/wcpa_categories/ (accessed 05.08.09).

49. Novelli, M., J. Barnes, and M. Humavindu (2006) 'The Other Side of the Ecotourism Coin Consumptive Tourism in Southern Africa', *Journal of Ecotourism* 5(1&2):62–79; and MacDonald, K. (2005) 'Global hunting grounds: Power, scale and ecology in the negotiation of conservation', *Cultural Geographies*, 12: 259–91.

CHAPTER 5: GUERRILLAS TO GORILLAS: BLOOD DIAMONDS AND COLTAN

1. Sandbrook, C. (2008) 'Putting leakage in its place: The significance of retained tourism revenue in the local context in rural Uganda', *Journal of International Development*, 22(1): 124–36; Adams, W. M. and M. Infield (2003) 'Who is on the gorilla's payroll? Claims on tourist revenue from a Ugandan National Park', *World Development*, 31(1): 177–90.

2. For further information see USAID's Central African Regional Programme for the Environment (CARPE) http://carpe.umd.edu/ (accessed 25.01.10); Dian Fossey Gorilla Fund International (DFGFI) http://www.gorillafund.org/ (accessed 25.01.10); and 'Impact of Congo Violence on Lowland Gorillas' *Scientific American* (23.07.10) at http://www.scientificamerican.com/article. cfm?id=congo-violence-and-gorillas.

3. Beswick, D. (2009) 'The challenge of warlordism to post-conflict state-building: The case of Laurent Nkunda in Eastern Congo', *The Round Table: The Commonwealth Journal of International Affairs*, 98(402): 333–46.

4. http://whc.unesco.org/en/list/137 (accessed 14.09.09); http://www.panda.org/ what_we_do/where_we_work/congo_basin_forests/wwf_solutions/congo_wild life_management/congo_gorillas/ (accessed 14.09.09).

5. http://gorilla.wildlifedirect.org/ (accessed 12.05.08).

6. 'Endangered Gorillas Threatened by Charcoal Trade', *The Times*, 12.12.09 at http://www.timesonline.co.uk/tol/news/world/africa/article6953669.ece (accessed 25.01.10); and see http://www.mercycorps.org/countries/drcongo/ 11789 (accessed 25.01.10); http://gorillacd.org/ (accessed 25.01.10).

7. 'UN experts to probe killing of highly endangered mountain gorillas in DR Congo' http://www.un.org/apps/news/story.asp?NewsID=23473&Cr= gorillas&Cr1= (accessed 05.05.08).

8. http://www.wcs.org/where-we-work/africa/congo.aspx (accessed 15.09.09).

9. http://www.oxfam.org/en/emergencies/congo (accessed 06.10.09).

10. For an overview of the changing nature of African international relations see Clapham, C. (1996) *Africa and the International System: The Politics of State Survival* (Cambridge, UK: Cambridge University Press).

11. For an excellent explanation of the genocide see Prunier, G. (1995) *The Rwanda Crisis: History of a Genocide* (New York: Columbia University Press); Pottier, J. (2002) *Re-Imagining Rwanda: Conflict, Survival and Disinformation in the Late Twentieth Century* (Cambridge, UK: Cambridge University Press).

12. Prunier, G. (2008) *Africa's World War: Congo, the Rwandan Genocide, and the Making of a Continental Catastrophe* (Oxford: Oxford University Press); see also Lemarchand, R. (2008) *The Dynamics of Violence in Central Africa* (Philadelphia: University of Pennsylvania Press).

13. 'Rwandan troops enter Congo', *Time Magazine*, 20.01.09.

14. Taylor, I. (2003) 'Conflict in Central Africa: clandestine networks and regional global configurations', *Review of African Political Economy*, 30 (95): 45–55; Clark, J. (ed.) (2002) *The African Stakes of the Congo War* (London: Palgrave).

15. United Nations (2001) Report of the Panel of Experts on the Illegal Exploitation of Natural Resources and Other Forms of Wealth of the Democratic Republic of the Congo. Available online http://www.un.org/News/dh/latest/drcongo.htm (accessed 15.09.09).

16. 'The human cost of mining in DR Congo', 02.09.09, BBC news at http://news.bbc.co.uk/1/hi/world/africa/8234583.stm (accessed 18.09.09); and

'Mineral firms "fuel Congo unrest" ', 21.07.09, at http://news.bbc.co.uk/1/hi/world/africa/8159977.stm (accessed 18.09.09).

17. 'Bisie killings show minerals at heart of Congo conflict', http://www.globalwitness.org/media_library_detail.php/801/en/bisie_killings_show_minerals_at_heart_of_congo_conflict (accessed 18.09.09).

18. http://www.globalwitness.org/pages/en/conflict_diamonds.html (accessed 18.09.09).

19. http://www.kimberleyprocess.com/ (accessed 16.09.09).

20. Kaldor, M. (2006) *New and Old Wars: Organized Violence in a Global Era* (Cambridge, UK: Polity, 2nd edn); see also Berdal, M. and D. M. Malone (2000) *Greed and Grievance: Economic Agendas in Civil Wars* (Boulder, CO: Lynne Rienner).

21. Marcus, R. R. and C. A. Kull (1999) 'Setting the stage: the politics of Madagascar's environmental efforts', *African Studies Quarterly* 3 (2): 1–7; Kull, C. A. (1996) 'The evolution of conservation efforts in Madagascar', *International Environmental Affairs*, 8(1): 50–86.

22. 'Madagascar titanium dioxide project approved', http://www.riotinto.com/media/media.aspx?id=896 (accessed 09.06.06); 'Madagascar's conservation conundrum', 11.04.04, http://news.bbc.co.uk/1/hi/world/africa/4433817.stm; http://www.fnxmining.com/ (accessed 26.10.09); 'South African companies have also invested, including Impala Platinum', see www.implats.co.za (accessed 26.10.09).

23. http://www.majescor.com/en/projects/other/madagascar.aspx (accessed 26.10.09); http://www.diamondfields.com/s/Madagascar.asp (accessed 26.10.09). Diamond Fields International also works in Liberia, Namibia and Greenland.

24. Horning, N. R. (2008) 'Strong support for weak performance: donor competition in Madagascar', *African Affairs*, 107(428): 405–31.

25. 'Down to business in Madagascar', 25.03.05, http://news.bbc.co.uk/1/hi/world/africa/4382733.stm; Duffy. R. (2007) 'Gemstone mining in Madagascar: transnational networks, criminalisation and global integration', *Journal of Modern African Studies*, 46(2):185–206.

26. Interview with staff member of the Institut de Gemmologie de Madagascar, Antananarivo, 18.03.04; 'Sapphires in the sand', *Focus* (August 2000). For further information on the World Bank reports on Madagascar and the sapphire sector see: http://web.worldbank.org/WBSITE/EXTERNAL/COUNTRIES/AFRICAEXT/MADAGASCAREXTN/0,,contentMDK:21344052~menuPK:356372~pagePK:141137~piPK:141127~theSitePK:356352,00.html (accessed 19.09.09).

27. 'Big hopes for Madagascan sapphires', *Financial Times*, 18.08.00.

28. Interview with staff member of the Institut de Gemmologie de Madagascar, Antananarivo, 18.03.04

29. Anonymised interviewees, Antananarivo, May 2004.

30. Duffy (2007) 'Gemstone mining in Madagascar'; see also Walsh, A. (2003) 'Hot money and daring consumption in a northern Malagasy sapphire mining town', *American Ethnologies*, 30 (2): 290–314.

31. Interview with Beatrice Olga Randrianarison, STD and AIDS Unit Head, Catholic Relief Services – Madagascar, Antananarivo, 30.03.04; also see http://www.catholicrelief.org/our_work/where_we_work/overseas/africa/madagascar/zalahy.cfm (accessed 15.08.04); interview with Parfait Randriamampianina, Director of Parks, ANGAP, Antananarivo, 21.08.01; pers. comm. Emison Jose, ANGAP guide, Isalo

National Park, 25.08.01. Also see, 'Prospectors and Poverty Mar an Island Paradise', *Financial Times*, 03.02.01.

32. Duffy (2007) 'Gemstone mining in Madagascar'.
33. Duffy (2007) 'Gemstone mining in Madagascar'.

CHAPTER 6: TOURIST SAVIOURS

1. 'Fears tourists helping illegal abalone trade', http://www.abc.net.au/news/stories/2004/04/28/1096422.htm (accessed 14.10.09).
2. http://www.operationcharm.org/news/20090508.jsp (accessed 26.01.10).
3. See Hanna, N. and S. Wells 'Saving the sea: tourism time bomb', *The New Internationalist*, 234, August 1992, at http://www.newint.org/issue234/bomb.htm (accessed 16.10.09).
4. 'Tourists targeted with hawksbill souvenirs', 06.08.09, at http://www.traffic.org/home/2009/8/6/tourists-targeted-with-hawksbill-shell-souvenirs.html (accessed 15.10.09).
5. 'Holidaymakers warned off illegal wildlife goods as 163,000 imports confiscated', *The Guardian* (UK), 16.08.07; and 'WWF warns holidaymakers over illegal souvenirs' at http://www.traffic.org/seizures/2007/8/16/wwf-warns-holidaymakers-over-illegal-souvenirs.html (accessed 30.10.09).
6. 'Belgians urged to leave future for souvenirs', at http://www.traffic.org/home/2008/6/13/belgians-urged-to-leave-a-future-for-souvenirs.html (accessed 30.10.09).
7. For more information see http://www.traffic.org/campaigns (accessed 30.10.09).
8. 'Turtles no longer turned to tourist souvenirs in Dominican Republic', 25.03.09 at http://www.traffic.org/home/2009/3/25/turtles-no-longer-turn-to-souvenirs-in-dominican-republic.html (accessed 15.10.09).
9. 'Wood carving industry hard on African forests', Yonge Nawe Environmental Action Group/Friends of the Earth Swaziland, 10.12.04, at http://www.yonge-nawe.com/04pressinformation/newsclips/10woodcarvingimpact101204.html (accessed 30.10.09).
10. CITES (2008) *National Wildlife Trade Policy Review: Madagascar*, available at http://www.cites.org/common/prog/policy/madagascar.pdf (accessed 30.10.09).
11. Pimm, S. 'The call to boycott Madagascar rosewood and ebony explained', *National Geographic*, 06.10.09 at http://blogs.nationalgeographic.com/blogs/news/chiefeditor/2009/10/madagascar-forest-crisis.html (accessed 15.10.09).
12. 'Bagpipes a threat to the environment', *The Scotsman*, 07.12.07, http://heritage.scotsman.com/heritage/Bagpipes-a-threat-to-the.3586303.jp (accessed 15.10.09).
13. http://www.fauna-flora.org/mpingo.php (accessed 15.10.09).
14. 'Ebony on the edge', *San Francisco Chronicle*, 18.09.09 (accessed 15.10.09).
15. McField, M. S. Wells and J. Gibson (eds) (1996), *State of the Coastal Zone Report Belize, 1995* (Belize City: Coastal Zone Management Programme/Government of Belize GOB), p. 114.
16. FAO (2002) *Report on Spiny Lobster Fisheries of Belize* at http://www.fao.org/docrep/006/Y4931B/y4931b07.htm (accessed 9.10.09).
17. Duffy, R. (2002) *A Trip Too Far: Ecotourism, Politics and Exploitation* (London: Earthscan), pp. 145–6.
18. 'Chiste grants licence to catch undersized lobster', *Amandala*, 13.7.97; 'Fisheries regulations amended', *San Pedro Sun*, 1.8.97; 'Editorial: lobster

farming – who to believe?', *San Pedro Sun*, 1.8.97; 'Taiwanese mafia business at South Water Caye', *Amandala*, 10.8.97.

19. 'Belling the Chinese cat', *Amandala*, 31.8.97.
20. 'Lobster season opens', *San Pedro Sun*, 27.6.97; 'September 30th marks end of closed season for the Queen Conch (Strombus gigas)', *BAS News*, 1998.
21. McField et al. (1996), *State of the Coastal Zone Report*, pp. 118–20.
22. Nietschmann, B. (1997) "Protecting Indigenous Coral Reefs and Sea Territories, Miskito Coast, RAAN, Nicaragua" in Stan Stevens (ed.), *Conservation Through Cultural Survival* (Washington, DC: Island Press), pp. 193–224.
23. For more information see Wildlife Friends of Thailand at http://www.wfft.org/ (accessed 19.10.09).
24. Lair, R. C. (2004) *Gone Astray: The Care and Management of The Asian Elephant in Domesticity* (4th edn) (Bangkok: FAO Regional Office for Asia and the Pacific).
25. Kontogeorgopoulos, N. (2009) 'Wildlife tourism in semi-captive settings: a case study of elephant camps in Northern Thailand', *Current Issues in Tourism* 12(5&6): 429–49; Cohen, E. (2008) *Explorations in Thai Tourism: Collected Case Studies* (Amsterdam: Elsevier Science).
26. Fennell, D. (1999) *Ecotourism: An Introduction* (London: Routledge); Cater, E. (2006) 'Ecotourism as a Western construct', *Journal of Ecotourism* 5(1 & 2): 23–39.
27. West, P. and J. G. Carrier (2004) 'Ecotourism and authenticity: getting away from it all?', *Current Anthropology*, 45(4): 483–98.
28. For more information see http://www.ceakumal.org/index.php (accessed 22.10.09).
29. http://www.belizegolf.cc/ (accessed 25.01.05).
30. http://www.cayechapel.com/ourisland.php (accessed 08.09.09).
31. http://www.cayechapel.com/ourisland.php (accessed 08.09.09). In 2009 the island was subdivided and individual estates and building plots were offered for sale after thirty years of individual ownership; they claim to be selling to the 'discriminating elite'.
32. Interview with representative of environmental NGO, Caye Caulker, 3.1.98; interview with Tour operator, Caye Caulker, 15.12.97. The controversial role of dredging is also mentioned in information about the resort – see http://www.ambergriscaye.com/ (accessed 25.01.05) – and remains a problem; see 'Is dredging driving away lobsters in Caye Caulker?', *San Pedro Daily*, 14.05.08.
33. Anonymous interviewee, Caye Caulker, January 1998. I also witnessed the activities of sand pirates while working there.
34. Coastal Zone Management Authority and Institute (undated) *Beach Erosion in Belize* (Belize City CZMA) p. 2. Sand mining is also listed as a threat to the Belize barrier reef by UNEP; see http://www.unep-wcmc.org/sites/wh/ reef.htm (accessed 25.01.05).
35. Anonymous interviewee, Caye Caulker, Belize, January 1998.
36. Anonymous interviewee, Caye Caulker, Belize, January 1998; and anonymous interviewee, San Pedro, Ambergis Caye, Belize, December 1997.
37. Duffy (2002) *A Trip Too Far*, pp. 127–54.
38. 'Belize by sea: much coast little guard', *Amandala*, 26.9.97. This was confirmed by a number of interviewees who shall remain anonymous.
39. 'Tory MP reported to ombudsman as Belize affair widens', *Guardian*, 22.7.99; 'Mr Ashcroft's patronage', *Guardian*, 22.7.99.

40. Ellis, G. (1995) 'Tourism and the Environment: Some Observations on Belize' in *Proceedings of the Second National Symposium on the State of the Belize Environment* (Belmopan: DoE), pp. 305–7.
41. Anonymous interviewee, Belize City, February 1998; also see Duffy (2002) *A Trip Too Far* (London: Earthscan) for further discussion.
42. Anonymous interviewees. The destruction of mangroves for tourism development in Placencia is also detailed in R. S. Pomeroy and T. Goetze (2003) *Belize Case Study: Marine Protected Areas Co-Managed by Friends of Nature* (Caribbean Conservation Association Report, in conjunction with DfID), p. 25.
43. Duffy, R. (2002) *A Trip Too Far: Ecotourism, Politics and Exploitation* (London: Earthscan) pp. 127–54.
44. 'A new class of trafficker', *Economist* 11–19th September, 1999, p. 66.
45. Belize by sea', *Amandala*, 26.9.97 and anonymous interviewees.
46. Bureau for International Narcotics and Law Enforcement (1996) *International Narcotics Control Strategy Report, 1997: Canada, Mexico and Central America*, p. 2.

CONCLUSION

1. http://www.panda.org/who_we_are/wwf_offices/cameroon/index.cfm?uProjectID=9F0725 (accessed 28.01.10).

BIBLIOGRAPHY

Adams, J. S. and McShane, T. O. (1992) *The Myth of Wild Africa: Conservation without Illusion* (Berkeley, CA: University of California Press)

Adams, W. M. (2004) *Against Extinction: The Story of Conservation* (London: Earthscan)

Adams, W. M. and Infield, M. (2003) 'Who is on the gorilla's payroll? Claims on tourist revenue from a Ugandan National Park', *World Development*, 31(1): 177–90

Adams, W. M. et al. (2004) 'Biodiversity conservation and the eradication of poverty?', *Science*, 306: 1146–9

Agrawal, A. (2005) 'Environmentality: Community, intimate government and the making of environmental subjects in Kumaon, India', *Current Anthropology*, 46(2): 161–90

Agrawal, A. and Gibson, C. C. (1999) 'Enchantment and disenchantment: The role of community in natural resource conservation', *World Development*, 27(4): 629–49

Agrawal, A. and Ostrom, E. (2001) 'Collective action, property rights, and decentralization in resource use in India and Nepal', *Politics and Society*, 29: 485–514

Ali, S. (2007) *Peace Parks, Conservation and Conflict Resolution* (Cambridge, MA: The MIT Press)

Baillie, J.E.M., Hilton-Taylor, C. and Stuart, S. N. (2004) *2004 IUCN Red List of Threatened Species: A Global Species Assessment* (Gland: IUCN)

Balmford, A. and Whitten, T. (2003) 'Who should pay for tropical conservation, and how could these costs be met?', *Oryx*, 37(2): 238–50

Barnett, M. (2002) *Eyewitness to a Genocide: The United Nations and Rwanda* (Ithaca, NY: Cornell University Press)

Barrow, E. and Fabricius, C. (2002) 'Do rural people really benefit from protected areas – rhetoric or reality?', *Parks*, 12: 67–79

de la Bedoyere, C. (2005) *No One Loved Gorillas More* (Bath: Palazzo Editions)

Berdal, M. and Malone, D. M. (2000) *Greed and Grievance: Economic Agendas in Civil Wars* (Boulder: Lynne Rienner)

Beswick, D. (2009) 'The challenge of warlordism to post-conflict state-building: The case of Laurent Nkunda in Eastern Congo', *The Round Table: The Commonwealth Journal of International Affairs*, 98(402): 333–46

Bolaane, M. (2004) 'The impact of game reserve policy on the River BaSarwa/Bushmen of Botswana', *Social Policy and Administration*, 38(4): 399–417

Bonner, R. (1993) *At the Hand of Man: Peril and Hope for Africa's Wildlife* (London: Simon and Schuster)

Borgerhoff-Mulder, M. and Coppolillo, P. (2005) *Conservation: Linking Ecology, Economics, and Culture* (Princeton, NJ: Princeton University Press).

Bowen-Jones, E. (2003) 'Bushmeat: traditional regulation or adaptation to market forces' in S. Oldfield (ed.) *The Trade in Wildlife: Regulation for Conservation* (London: Earthscan), pp. 132–45

Bradley Martin, E. and Bradley Martin, C. (1982) *Run Rhino Run* (London: Chatto and Windus)

Brandon, K., Redford, K. H. and Sanderson, S. E. (1998) *Parks in Peril: People, Politics and Protected Areas* (Washington D.C.: The Nature Conservancy)

Brechin, S.R. et al. (2003) *Contested Nature. Promoting International Biodiversity with Social Justice in the Twenty-first Century* (Albany, NY: State University of New York Press)

Broad, S., Mulliken, T. and Roe, D. (2003) 'The nature and extent of legal and illegal trade in wildlife' in S. Oldfield (ed.) *The Trade in Wildlife: Regulation for Conservation* (London: Earthscan), pp. 3–22

Brockington, D. (2009) *Celebrity and the Environment: Fame, Wealth and Power in Conservation* (London: Zed Books)

Brockington, D., Duffy, R. and Igoe, J. (2008) *Nature Unbound: Conservation, Capitalism and the Future of Protected Areas* (London: Earthscan)

Brockington, D. and Igoe, J. (2006) 'Eviction for conservation: A global overview', *Conservation and Society*, 4(3): 424–70

Brooks, T. M. et al. (2002) 'Habitat loss and extinction in the hotspots of biodiversity', *Conservation Biology*, 16(4): 909–23

Brooks, T. M. et al. (2006) 'Global biodiversity conservation priorities', *Science*, 313(5783): 58–61

Brosius, J. P. (2006) 'Common round between anthropology and conservation biology', *Conservation Biology*, 20(3): 683–5

Brosius, J. P., Tsing, A. L. and Zerner, C. (eds) (2005) *Communities and Conservation: Histories and Politics of Community-based Natural Resource Management* (Walnut Creek, CA: Altamira)

Bulbeck, C. (2004) *Facing the Wild: Ecotourism, Conservation and Animal Encounters* (London: Earthscan)

Büscher, B. (2010) 'Seeking telos in the "transfrontier": Neoliberalism and the transcending of community conservation in Southern Africa', *Environment and Planning A* 42(3): 644–60

Büscher, B. (2010) 'Anti-politics as political strategy: Neoliberalism and transfrontier conservation in Southern Africa', *Development and Change*, 41(1): 29–51

Bibliography

Büscher, B. and Dressler, W. (2007) 'Linking neoprotectionism and environmental governance: On the rapidly increasing tensions between actors in environment–development nexus', *Conservation and Society*, 5(4): 586–611

Butchart, S. H. M., Stattersfield, A. J. and Collar, N. J. (2006b) 'How many bird extinctions have we prevented?', *Oryx*, 40(3): 266–78

Carruthers, J. (1995) *The Kruger National Park: A Social and Political History* (Pietermaritzburg: University of Natal Press)

Cater, E. (2006) 'Ecotourism as a Western construct', *Journal of Ecotourism*, 5(1&2): 23–39

Chapin, M. (2004) 'A challenge to conservationists', *World Watch Magazine*, 17(6):17–31

Child, B. (ed.) (2004) *Parks in Transition: Biodiversity, Rural Development and the Bottom Line* (London: Earthscan)

Clapham, C. (1996) *Africa and the International System: The Politics of State Survival* (Cambridge: Cambridge University Press)

Clark, J. (ed.) (2002) *The African Stakes of the Congo War* (London: Palgrave)

Cohen, E. (2008) *Explorations in Thai Tourism: Collected Case Studies* (Amsterdam: Elsevier Science)

Cooney, R. and Jepson, P. (2006) 'The international wild bird trade: What's wrong with blanket bans?', *Oryx*, 40(1): 18–23

Corson, C. (forthcoming 2010) 'Shifting environmental governance in a neoliberal world: The politics of US(aid) for conservation', *Antipode: A Journal of Radical Geography*

Dallaire, R. (2003) *Shake Hands with the Devil: The Failure of Humanity in Rwanda* (Toronto: Random House)

Demerrit, D. (2006) 'Science studies, climate change and the prospects for a constructivist critique', *Economy and Society*, 35(3): 453–79

Dickson, B. (2003) 'What is the goal of regulating the wildlife trade? Is regulation a good way to achieve this goal?' in S. Oldfield, (ed) *The Trade in Wildlife: Regulation for Conservation* (London: Earthscan), pp. 23–32

Dowie, M. (2009) *Conservation Refugees: The Hundred-Year Conflict between Global Conservation and Native Peoples* (Cambridge: MIT Press)

Dressler, W. and Büscher, B. (2008) 'Market triumphalism and the so-called CBNRM "crisis" at the South African section of the Great Limpopo Transfrontier Park', *Geoforum*, 39(1): 452–65

Duffy, R. (1999) 'The role and limitations of state coercion: anti-poaching policies in Zimbabwe', *Journal of Contemporary African Studies*, 17: 97–121

Duffy, R. (2000) *Killing for Conservation: The Politics of Wildlife in Zimbabwe* (Oxford: James Currey/ Indiana: Indiana University Press, African Issues Series)

Duffy, R. (2002) *A Trip Too Far: Ecotourism, Politics and Exploitation* (London: Earthscan)

Duffy, R. (2006) 'Global governance and environmental management: The politics of transfrontier conservation areas in Southern Africa', *Political Geography*, 25(1): 89–112

Duffy. R. (2007) 'Gemstone mining in Madagascar: transnational networks, criminalisation and global integration', *Journal of Modern African Studies*, 46(2): 185–206

Duffy, R. (2008) 'Ivory Trade: Enforceability, effectiveness and ethical concerns' in C. Wemmer and K. Christen (eds.) *Elephants and Ethics: The Morality of Coexistence* (Johns Hopkins Press), pp. 451–467

Dzingirai, V. (2003) 'The new scramble for the African countryside', *Development and Change*, 34(2): 243–63

Ehrlich, P.R. and Wilson, E. O. (1991) 'Biodiversity Studies: Science and Policy' *Science*, 253: 758–62

EIA (1989) *A System of Extinction: the African Elephant Disaster* (London: EIA)

EIA (1994) *CITES: Enforcement Not Extinction* (London: EIA)

EIA (2008) *Skin Deep* (London: EIA)

EIA, David Shepherd Foundation and Tusk Force (1993) *Taiwan Kills Rhinos with Your Money: Why You Should Boycott Goods Made in Taiwan* (London: EIA)

Ellis, G. (1995) 'Tourism and the environment: Some observations on Belize', in *Proceedings of the Second National Symposium on the State of the Belize Environment* (Belmopan: DoE)

Engler, M. and Parry-Jones, R. (2007) *Opportunity or Threat: The Role of the European Union in Global Wildlife Trade* (Brussels: TRAFFIC-Europe)

Fabricius, C. and de Wet, C. (2002) 'The influence of forced removals and land restitution on conservation in South Africa', in D. Chatty and M. Colchester (eds) *Conservation and Mobile Indigenous Peoples. Displacement, Forced Settlement, and Sustainable Development* (New York: Berghahn Books)

Fennell, D. (1999) *Ecotourism: An Introduction* (London: Routledge)

Ferguson, J. (2006) *Global Shadows: Africa in the Neoliberal World* (Durham NC: Duke University Press)

Forsyth, T. and Walker, A. (2008) *Forest Guardians, Forest Destroyers: The Politics of Environmental Knowledge in Northern Thailand* (Seattle, WA: University of Washington Press)

Fossey, D. (1983) *Gorillas in the Mist* (Boston, MA: Houghton Mifflin)

Garland, E. (2008) 'The elephant in the room: Confronting the colonial character of wildlife conservation in Africa', *African Studies Review*, 51(3): 51–74

Goldman, M., (2001) 'Constructing an environmental state: Eco-governmentality and other practices of a "green" World Bank', *Social Problems*, 48: 499–523

Goldman, M. (2005) *Imperial Nature: The World Bank and Struggles for Social Justice in the Age of Globalization* (New Haven, CT: Yale University Press)

Halpern, B. S. et al. (2006) 'Gaps and mismatches between global conservation priorities and spending', *Conservation Biology*, 20(1): 56–64

Hayes, H. T. P (1989) *The Dark Romance of Dian Fossey* (London: Simon and Schuster)

Hayes, T. M. (2006) 'Parks, people, and forest protection: An institutional assessment of the effectiveness of protected areas', *World Development*, 34(12): 2064–75

Hayes, T. M. and Ostrom, E. (2005) 'Conserving the world's forests: Are protected areas the only way?', *Indiana Law Review*, 38: 595–617

Honey, M. (1999) *Ecotourism and Sustainable Development. Who Owns Paradise?* (Washington D.C.: Island Press)

Hoon, P. (2004) 'Impersonal markets and personal communities? Wildlife, conservation, and development in Botswana', *Journal of International Wildlife Law and Policy*, 7:143–60

Horning, N. R. (2008) 'Strong support for weak performance: Donor competition in Madagascar', *African Affairs*, 107(428): 405–31

Hulme, D. and Edwards, M. (1997) *NGOs, States and Donors: Too Close for Comfort?* (London: Macmillan)

Hulme, D. and Murphee, M. (eds) (2001) *African Wildlife and Livelihoods: The Promise and Performance of Community Conservation* (Oxford: James Currey)

Bibliography

Hutton, J., Adams, W. M. and Murombedzi, J. C. (2005) 'Back to the barriers? Changing narratives in biodiversity conservation', *Forum for Development Studies*, 32(2): 341–70

Hutton, J. and Dickson, B. (2000) *Endangered Species, Threatened Convention: The Past, Present and Future of CITES* (London: Earthscan)

IFAW (2008) *Killing with Keystrokes: An Investigation of the Illegal Wildlife Trade on the World Wide Web* (Brussels: IFAW)

Igoe, J. and Fortwrangler, C. (2007) 'Whither communities and conservation?' *International Journal of Biodiversity Science and Management*, 3: 65–76

IUCN (1980) *The International Trade in Rhinoceros Products*, report prepared by E. B. Martin for IUCN and WWF (Gland, Switzerland: IUCN/WWF)

IUCN (1992) *Analysis of Proposals to Amend the CITES Appendices*, prepared by the IUCN Species Survival Commission Trade Specialist Group, TRAFFIC and WCMC for the 8th Meeting of the Conference of the Parties, Kyoto Japan 2–13 March 1992 (Gland, Switzerland: IUCN)

Ivory Trade Review Group (1989) *The Ivory Trade and the Future of the African Elephant*, vol. 1 (summary and conclusions), report prepared for the 7th CITES, Conference of the Parties, Lausanne, October 1989 (Oxford: ITRG)

Jacoby, K. (2003) *Crimes Against Nature Squatters, Poachers, Thieves, and the Hidden History of American Conservation* (Berkeley, CA: University of California Press)

Kaldor, M. (2006) *New and Old Wars: Organised Violence in a Global Era* (Cambridge: Polity, second edition)

Kapp, C. (2008) 'Crystal meth boom adds to South Africa's health challenges', *The Lancet*, 371(9608): 193–4

King, B. (forthcoming) 'Conservation geographies in Sub-Saharan Africa: The politics of national parks, community conservation and peace parks', *Geography Compass*

King, B., and S. Wilcox (2008) 'Peace parks and jaguar trails: Transboundary conservation in a globalizing world', *GeoJournal*, 71(4): 221–31

Knapp, A., Kitschke, C. and von Meibom, S. (eds.) (2006) *Proceedings of the International Sturgeon Enforcement Workshop to Combat Illegal Trade in Caviar* (Brussels: TRAFFIC-Europe)

Kontogeorgopoulos, N. (2009) Wildlife tourism in semi-captive settings: A case study of elephant camps in Northern Thailand', *Current Issues in Tourism*, 12(5&6): 429–49

Kull, C. A. (1996) 'The evolution of conservation efforts in Madagascar', *International Environmental Affairs*, 8(1): 50–86

Kull, C. A. (2002) 'Madagascar's burning issue: the persistent conflict over fire', *Environment*, 44(3): 8–19

Lair, R. C. (2004) *Gone Astray: The Care and Management of the Asian Elephant in Domesticity* (fourth edition), Bangkok: FAO Regional Office for Asia and the Pacific

Leader-Williams, N. (2003) 'Regulation and protection: successes and failure in rhinoceros conservation' in S. Oldfield (ed.) *The Trade in Wildlife: Regulation for Conservation* (London: Earthscan), pp. 89–99

Leader-Williams, N., Milledge, S., Adcock, K., Brooks, M., Conway, A., Knight, M., Mainka, S., Martin, E. B. and Teferi, T. (2005) 'Trophy hunting of black rhino Diceros bicornis: Proposals to ensure its future sustainability', *Journal of International Wildlife Law and Policy*, 8: 1–11

Leakey, R. (2001) *Wildlife Wars: My Battle to Save Kenya's Elephants* (London: Pan)

Lemarchand, R. (2008) *The Dynamics of Violence in Central Africa* (Philadelphia: University of Pennsylvania Press)

Li, T. (2007) *The Will to Improve: Governmentality, Development, and the Practice of Politics* (Durham: Duke University Press)

Litfin, K. (1994) *Ozone Discourses: Science and Politics in Global Environmental Co-Operation* (New York: Columbia University Press)

MacDonald, K. (2005) 'Global hunting grounds: Power, scale and ecology in the negotiation of conservation', *Cultural Geographies*, 12: 259–91

MacKenzie, J. M. (1988) *Empire of Nature: Hunting Conservation and British Imperialism* (Manchester: Manchester University Press)

Marcus, R. R. and Kull, C. A. (1999) 'Setting the stage: The politics of Madagascar's environmental efforts', *African Studies Quarterly*, 3(2): 1–7

Mbaiwa, J. (2004) 'The socio-cultural impacts of tourism development in the Okavango Delta, Botswana', *Journal of Tourism and Cultural Change*, 2(2): 163–85

Mbaiwa, J. (2008) 'The realities of ecotourism development in Botswana', in A. Spenceley (ed.) *Responsible Tourism: Critical Issues for Conservation and Development* (London: Earthscan), pp. 205–23

McField, M., Wells, S. and Gibson. J. (eds) (1996) *State of the Coastal Zone Report Belize, 1995* (Coastal Zone Management Programme/Government of Belize GOB: Belize City)

Melvern, L. (2009) *A People Betrayed: The Role of the West in Rwanda's Genocide* (London: Zed Books, second edition)

Milliken, T. (1994) 'Current trade in, demand for and stockpiles of ivory in range and consumer states', paper presented at *The African Elephant in the Context of CITES*, conference held in Kasane, Botswana 19–23 September 1994 (UK Department of Environment)

Mills, A. (1997) *Rhinoceros Horn and Tiger Bone in China: An Investigation of Trade since the 1993 Ban* (Cambridge: TRAFFIC)

Mittermeier, R. A., Myers, N., Robles-Gil, P. and Mittermeier, C. G. (eds) (1999) *Hotspots: Earth's Biologically Richest and Most Endangered Terrestrial Ecoregions* (CEMEX/Agrupación Sierra Madre, Mexico City)

Mittermeier, R. A. et al (2004) *Hotspots Revisited: Earth's Biologically Richest and Most Endangered Terrestrial Ecoregions* (CEMEX/Agrupación Sierra Madre, Mexico City)

Moore, L. (2009) 'Conservation heroes versus environmental villains: Perceiving elephants in Caprivi, Namibia', *Human Ecology*, 25: 147–69

Morris, S. D. (1999) 'Corruption and the Mexican political system: continuity and change', *Third World Quarterly*, 20(3)

Mukamuri, B. B, Manjengwa, J. M. and Anstey, S. (eds) (2009) *Beyond Proprietorship: Murphree's Laws on Community-Based Natural Resource Management in Southern Africa* (Harare: Weaver Press/Ottawa: International Development Research Centre)

Murombedzi, J. (2001) 'Why wildlife conservation has not economically benefited communities in Africa', in D. Hulme and M. Murphree (eds) *African Wildlife and Livelihoods* (Oxford: James Currey), pp. 208–26

Myers, N. (1988) 'Threatened biotas: "Hot spots" in tropical forests', *The Environmentalist*, 8(3): 187–208

Myers, N. (1990) 'The biodiversity challenge: expanded hot-spots analysis', *The Environmentalist*, 10(4): 243–56

Neumann, R. P. (1998) *Imposing Wilderness: Struggles over Livelihood and Nature Preservation in Africa* (Berkeley, CA: University of California Press)

Neumann, R. P. (2001) 'Disciplining peasants in Tanzania: From state violence to self surveillance in wildlife conservation', in N. L. Peluso and M. Watts (eds) *Violent Environments* (Ithaca/London: Cornell University Press), pp. 305–27

Bibliography

Neumann, R. P. (2004) 'Moral and discursive geographies in the war for biodiversity in Africa', *Political Geography*, 23: 813–37

Nietschmann, B., (1997) 'Protecting indigenous coral reefs and sea territories, Miskito Coast, RAAN, Nicaragua' in S. Stevens (ed.) *Conservation Through Cultural Survival* (Washington D. C.: Island Press), pp. 193–224

Nordstrom, C. (2004) *The Shadows of War: Violence, Power, and International Profiteering in the Twenty-First Century* (Berkeley, CA: University of California Press)

Novelli, M., Barnes, J. and Humavindu, M. (2006) 'The other side of the ecotourism coin: Consumptive tourism in Southern Africa', *Journal of Ecotourism*, 5(1&2): 62–79

Nowell, K. (2000) *Far from a Cure: The Tiger Trade Revisited* (Cambridge: TRAFFIC International)

Nowell, K. (2007) *Asian Big Cat Conservation and Trade Control in Selected Range States: Evaluating Implementation and Effectiveness of CITES Recommendations* (Cambridge: TRAFFIC International)

Nowell, K. and Xu, L. (2007) *Taming the Tiger Trade: China's Markets for Wild and Captive Tiger Products since the 1993 Domestic Trade Ban* (TRAFFIC East Asia)

Oates, J. F. (1999) *Myth and Reality in the Rainforest: How Conservation Strategies Are Failing West Africa* (Berkeley: University of California Press)

Oldfield, S. (ed.) (2003) *The Trade in Wildlife: Regulation for Conservation* (London: Earthscan)

Ostrom, E. and Nagendra, H. (2006) 'Insights on linking forests, trees and people from the air, on the ground, and in the laboratory', *Proceedings of the National Academy of Sciences of the United States of America*, 103(51): 19224–31

Patullo, P. (1996) *Last Resorts: The Cost of Tourism in the Caribbean* (London: Cassell)

Pawliczek, M. and Mehta, H. (2008) 'Ecotourism in Madagascar: How a sleeping beauty is finally awakening' in A. Spenceley (ed.) *Responsible Tourism: Critical Issues for Conservation and Development* (London: Earthscan), pp. 41–68

Pearce, F. (2005) 'Laird of Africa', *New Scientist*, 13 August 2005: 48–50

Peluso, N. L. 'Coercing conservation?: The politics of state resource control', *Global Environmental Change*, 3(2): 199–218

Peluso, N. L. and Watts, M. (eds) (2001) *Violent Environments* (Ithaca: Cornell University Press)

Pimm, S. L., et al. (1995) 'The future of biodiversity', *Science*, 269: 347–50

Pimm, S. L., and Raven, P. (2000) 'Extinction by numbers', *Nature*, 403: 843–5

Pomeroy, R. S. and Goetze, T. (2003) *Belize Case Study: Marine Protected Areas Co-Managed by Friends of Nature* (Caribbean Conservation Association Report, in conjunction with DfID)

Pottier, J. (2002) *Reimagining Rwanda: Conflict, Survival and Disinformation in the Late Twentieth Century* (Cambridge: Cambridge University Press)

Prunier, G. (1995) *The Rwanda Crisis: History of a Genocide* (New York: Columbia University Press)

Prunier, G. (2008) *Africa's World War: Congo, the Rwandan Genocide, and the Making of a Continental Catastrophe* (Oxford: Oxford University Press)

Ramutsindela, M. (2007) *Transfrontier Conservation in Africa: At the Confluence of Capital, Politics and Nature* (Wallingford: CABI)

Randrianja, S. and Ellis, S. (2009) *Madagascar: A Short History* (London: Hurst and Company)

Ranganathan, J. et al. (2008) 'Where can tigers persist in the future? A landscape-scale density-based population model for the Indian subcontinent', *Biological Conservation*, 141: 67–77

Redford, K. H. and Adams, W. A. (2009) 'Payment for ecosystem services and the challenge of saving nature', *Conservation Biology*, 23(4): 785–7

Reeve, R. and Ellis, S. (1995) 'An insider's account of the South African security force's role in the ivory trade', *Journal of Contemporary African Studies*, 13: 222–43

Reid, H. (2001) 'Contractual national parks and the Makuleke community', *Human Ecology*, 29(2): 135–55

Ribot, J. C. (2004) *Waiting for Democracy: The Politics of Choice in Natural Resource Decentralisation* (Washington DC: World Resources Institute)

Rivalan, P., Delmas, V., Angulo, E., Bull, L. S., Hall, R. J., Courchamp, F., Rosser, A. M. and Leader-Williams, N. (2007) 'Can bans stimulate wildlife trade?', *Nature*, 447: 529–30

Robbins, P., McSweeney, K., Chhangani, A. K. and Rice, J. L. (2009) 'Conservation as it is: Illicit resource use in a wildlife reserve in India', *Human Ecology*, 37(5): 559–75

Roe, D. (2006) 'Blanket bans: conservation or imperialism? A response to Cooney & Jepson', *Oryx*, 40(1): 27–8

Roe, D. (2008) *Trading Nature: A Report with Case Studies, on the Contribution of Wildlife Trade Management to Sustainable Livelihoods and the Millennium Development Goals* (Cambridge: TRAFFIC International and WWF International)

Sand, P. H. (2009) *The United States and Britain in Diego Garcia: The Future of a Controversial Base* (London: Palgrave/Macmillan)

Sandbrook, C. (2008) 'Putting leakage in its place: The significance of retained tourism revenue in the local context in rural Uganda', *Journal of International Development*, 22(1): 124–36

Sanderson, S. E. and Redford, K. H. (2003) 'Contested relationships between biodiversity conservation and poverty alleviation?', *Oryx*, 37(4): 1–2

Schmidt-Soltau, K. and Brockington, D. (2007) 'Protected areas and resettlement: What scope for voluntary relocation', *World Development* 35(12): 2182–2202

Scholfield, K. and Brockington, D. (forthcoming 2010) 'Actually existing conservation: Understanding conservation NGOs in sub-Saharan Africa', *Antipode: A Journal of Radical Geography*

Steinberg, J. (2005) *The Illicit Abalone Trade in South Africa* (Johannesburg: Institute for Security Studies)

Stonich, S. C., Sorensen, J. H. and Hundt, A. (1995) 'Ethnicity, class and gender in tourism development: The case of the Bay Islands, Honduras', *Journal of Sustainable Tourism*, 3(1): 1–28

Taylor, I. (2003) 'Conflict in Central Africa: Clandestine networks and regional global configurations', *Review of African Political Economy*, 30(95): 45–55

Terborgh, J. (1999) *Requiem for Nature* (Washington D. C.: Island Press)

Thornton, A. and Currey, D. (1991) *To Save an Elephant* (London: Bantam Books)

TRAFFIC (2008) *What's Driving the Wildlife Trade? A Review of Expert Opinion on Economic and Social Drivers of the Wildlife Trade and Trade Control Efforts in Cambodia, Indonesia, Lao PDR and Vietnam* (Cambridge: TRAFFIC International/Washington DC: World Bank)

Tsing, A. L. (2004) *Friction: An Ethnography of Global Connection* (Princeton, NJ: Princeton University Press)

Vine, D.S. (2009) *Island of Shame: The Secret History of the US Military Base on Diego Garcia* (Princeton, NJ: Princeton University Press)

Vira, B. and Adams, W.M. (2009) 'Ecosystem services and conservation strategy: Beware the silver bullet', *Conservation Letters*, 2: 158–62

Bibliography

Walsh, A. (2003) 'Hot money and daring consumption in a Northern Malagasy sapphire mining town', *American Ethnologist*, 30(2): 290–314

Weber, B. and Veder, A. (2001) *In the Kingdom of Gorillas: Fragile Species in a Dangerous Land* (London: Simon and Schuster)

Wels, H. (2003) *Private Wildlife Conservation in Zimbabwe: Joint Ventures and Reciprocity* (Leiden: Brill Academic Publishers)

Wells, M., Brandon, K. and Hannah, L. (1992) *People and Parks: Linking Protected Area Management with Local Communities* (Washington D. C.: The World Bank)

West, P. and Carrier, J. G. (2004) 'Ecotourism and authenticity: Getting away from it all?', *Current Anthropology*, 45(4): 483–98

Western, D. and Wright, R. (eds) (1994) *Natural Connections: Perspectives on Community Based Conservation* (Washington D. C.: Island Press)

Wilkie, D. S. et al. (2006) 'Parks and people: Assessing the human welfare effects of establishing protected areas for biodiversity conservation', *Conservation Biology*, 20(1): 247–9

Williamson, D. F. and Henry, L. A. (2008) *Paper Tigers? The Role of the US Captive Tiger Population in the Trade in Tiger Parts* (Washington D. C.: TRAFFIC US)

Wilson, E. O. (ed.) (1988) *Biodiversity* (Washington D. C.: National Academy of Sciences/Smithsonian Institution)

Wilson E. O. (1992) *Diversity of Life* (Cambridge, MA: Harvard University Press)

Wilson, K. A. et al. (2006) 'Prioritizing global conservation efforts', *Nature*, 440(7082): 337–40

Wilson, K. A. et al (2007) 'Conserving biodiversity efficiently: What to do, where, and when', *PLoS Biology*, 5(9): 1850–61

Wolmer, W. (2006) *From Wilderness to Farm Invasions: Conservation and Development in Zimbabwe's Southeast Lowveld* (Oxford: James Currey)

Zimmerer, K. S. (2006) 'Cultural ecology at the interface with political ecology: The new geographies of environmental conservation and globalisation', *Progress in Human Geography*, 30(1): 63–78

INDEX

abalone, 23–6, 44, 118, 187–8
  linked to drugs (tik), 25–6
Afghanistan, 28
African blackwood *see* ebony wood
African Parks Foundation, 58–9
African Rainforest and River
  Conservation (ARRC), 107
African Wildlife Foundation, 49, 94,
  126–7
African Wildlife Leadership
  Foundation, 135
Aldabran tortoise, 85
Amazon, 64
American Association of Zoos and
  Aquariums, 93
ANGAP, 182, 183
Angola, 95, 165, 173; *see also* UNITA
Animal Welfare Institute, 128
Attenborough, Sir David, 64, 156
Australia, 37, 137, 187–8

Bangladesh, 27
Belize, 8, 199–202, 209–215
  corruption, 213–14
Belize Ecotourism Association, 212
Bhutan, 27
biodiversity, 12, 61, 178
  and national parks, 56–7, 108

and NGOs, 72–3, 75
meanings of, 51–3, 221.
'war for' 96, 99, 106; *see also*
  poaching
birds' nests, 15–16
bonobos, 92
Bosack and Kruger Charitable
  Foundation, 35
Botswana, 8, 60–1, 79–80, 130, 131,
  140, 145, 169
BP, 68, 69
bushmeat, 34, 68, 84, 91–3
Bushmeat Crisis Task Force, 92, 93
Bwindi Impenetrable Forest 155,
Bwindi Impenetrable National Park,
  155–6

Cambodia, 27
CAMPFIRE, 142; *see also* CBNRM
Catholic Relief Services, 179–80
caviar trade, 21–3, 34, 44, 218
Caye Chapel, 209–11
Central African Republic, 165
Central Kalahari Game Reserve, 60
Centro Ecological de Akumal,
  208–9
Chagos Archipelago, 5–6
charcoal trade, 161–2

China, 16, 27, 28, 29, 36, 40–3, 45, 89,
    94, 117, 168, 187
climate change, 69
cockling, 37–43, 218
  deaths of Chinese cocklers, 39–40
Cold War, 64–5, 106, 113, 165–6, 173
colonialism, 5, 84–6, 123–5
coltan, 13, 160–1,164, 168, 185
Community Based Natural Resource
    Management (CBNRM), 97–9,
    133; see also CAMPFIRE
Congolese Institute for Nature
    Conservation, 161
Conservation International (CI),
    1, 2, 6, 7, 53, 66, 69, 73–4, 80,
    92, 163
  Global Conservation Fund, 75
Convention on Biological Diversity,
    52, 53, 68, 76
Convention on the International Trade
    in Endangered Species (CITES)
    45–51, 76
  appendices, 21, 24, 36, 47–8, 137,
    139, 191, 196, 197
  Conference of the Parties, 122
  Elephant Trade Information System
    (ETIS), 148
  establishment of, 46
  membership, 46
  National Management Authority, 47
  NGOs, 68, 136
  scientific authority, 47
  split listings, 21, 47, 137, 138
  and timber trading timber, 194–5
  Western bias, 50
  and wildlife trading, 28–9, 31, 45,
    89, 115, 120–3
  see also ivory; rhinos; tigers
Cool Earth, 63–4
curio trade see seashells; wooden
    carvings

Dalai Lama, 28, 90
David Shepherd Wildlife Foundation
    (DSWF), 34
Democratic Republic of Congo
    (DRC), 10, 13, 92, 104, 156–74,
    176, 179
Department of Environment, Food
    and Rural Affairs (DEFRA), 32, 33
diamonds, 10, 26, 107, 158, 160, 164,
    168–71, 174, 185
  blood diamonds, 171–3
  see also Kimberley Process

Dian Fossey Gorilla Fund
    International (DFGI),
    92, 163
Diego Garcia, 5
dodo, 85
donors, 3, 135
DreamWorks, 68
drug trafficking, 118, 201–2, 211–15,
    216
Durrell Wildlife Conservation Trust,
    4, 195

Earthwatch, 7
eBay, 35–7
ebony wood, 188, 195–8; see also
    wooden carvings
ecotourism, 7, 68, 70, 161, 181, 206
Ehrlich, Paul, 2
elephants, 4, 46, 47, 54, 149–50, 156
  culling, 130, 143–5
  elephant camps, 205
  growing populations, 129–34, 220
  Problem Animal Control (PAC),
    143, 145–7
  Thai Elephant Conservation Centre,
    205
  trained elephants, 203–5
  translocation, 147–8
  Year of the Elephant (1988), 127
  see also ivory
Eliasch, Johan, 63–4
Environmental Investigation Agency
    (EIA), 28, 29, 81, 88, 121
  and the ivory trade ban, 127–8
epistemic communities, 48–9, 67–8
Ethiopia, 8, 58–9
extinction crisis, 1–2, 4, 217
Exxon Mobil, 3, 27, 69

fishing, 20–1
Flora and Fauna International, 197
Ford, Harrison, 34, 52, 65, 69
Fossey, Dian, 100, 101–3, 156
Friends of the Earth, 7
Frankfurt Zoological Society, 57,
    160

Galapagos, 85
gibbons, 203
Global Witness, 171
Gordon and Betty Moore Foundation,
    67
gorillas, 2, 3, 10, 14, 100, 101–3, 172,
    174, 184, 185, 218

Eastern lowland gorillas, 156,
159,161
gorilla murders in DRC, 156–64
mountain gorillas, 46, 156
tracking permits, 156, 216
Western lowland gorillas, 159
Great Limpopo Transfrontier Park, 62
Greenpeace, 127
Grzimek, Bernard, 57

HM Customs and Excise, 32–3
Hong Kong, 23, 25, 119, 137
Humane Society of the US (HSUS),
145
hunting
Professional Hunting Association of
South Africa (PHASA), 122, 146
sport hunting, 80, 84–6, 87, 121–3,
143, 146–7, 149
subsistence hunting, 85–8, 222
see also poaching

India, 27, 28, 43, 47, 60–1, 89–90
Indonesia, 27, 45, 71–3
Integrated Conservation and
Development Project (ICDP), 72
International Biological Diversity
Day, 53
International Conservation Caucus
Foundation, 69
International Fund for Animal
Welfare (IFAW), 30, 34, 35, 36,
81, 121, 145
International Rhino Foundation, 115
International Union for the
Conservation of Nature (IUCN),
2, 4, 6, 19, 27, 28, 70, 120, 149
IUCN African Elephant Specialist
Group, 121, 135
IUCN African Rhino Specialist
Group, 135
IUCN Redlist, 2, 27
IUCN Species Survival Commission,
121
International Year of Biodiversity, 53
Interpol, 33, 35, 218
ivory, 192, 218
history of the ivory trade, 123–5
illegal trading, 10, 13, 18, 36, 50,
101, 124, 134, 152
ivory bonfire, 135–6
legal trading, 132
poaching, 4, 84, 86, 94, 126
reopening the trade, 148–50, 153

split between east and southern
Africa, 134, 139–43
stockpiles, 143, 144–8, 220
trade ban, 113, 126–39, 152–4,
219–20
uses of, 85, 123–5
Ivory Trade Review Group (ITRG),
136

Japan, 94, 125, 134

Kabila, Joseph, 167; see also Democratic
Republic of Congo
Kabila, Laurent, 167, 168; see also
Democratic Republic of Congo
Kagame, Paul, 167; see also Rwanda
Kahuzi-Biega National Park, 161
Kenya
elephant populations, 126
and ivory debate, 122–3, 128, 134–6
tourism, 152, 188, 216
war on poachers, 82, 96, 100–1, 104
wooden carving industry, 197
Kenya Wildlife Services (KWS), 82,
100, 134–6
Kew Gardens, 5, 7
Kimberley Process Certification
Scheme, 171–2, 179
Kruger National Park, 54–5, 63, 107,
145, 152
Kumbhalgarh Wildlife Sanctuary, 60–2
Kumleben Commission, 95

Lamanai Declaration, 199
Laos, 27
Leakey, Dr Richard, 100, 104, 134–6,
148, 160; see also Wildlife Direct
lemurs, 176–8, 181, 184, 195, 218
Liwonde National Park, 82
lobsters, 198–200
farming, 200–1
illegal fishing, 201–2
London Zoo, 5
lorises, 202

Maasai, 57
Macao, 119
Madagascar, 5, 6–7, 8, 15–17, 18, 51,
68, 218
Durban Vision Initiative, 73–6
Ilakaka, 176–84, 185,
Isalo National Park, 176–8, 181–2
Madagascar Fauna Group, 195
sapphire mining, 174–85

Madagascar (*cont.*)
  wooden curio trade, 194–6
  see also Qit Minerals Madagascar;
    Ravalomanana, Marc
Mafia, 35, 118
Malawi, 82, 194, 196, 197
Malaysia, 27, 28
Maldives, 190–1
Marine Conservation Society, 5
Marine Resources Assessment Group
  (MRAG), 20
Mauritius, 4, 85
Mauritius kestrel, 4
McDonalds, 68
Mekong River Basin Development
  Programme, 70
Mexico, 8, 45, 201, 202, 207–9
Mineral Resources Governance
  Project, 175, 176
Missouri Botanical Gardens, 7
Moi, Daniel Arap, 82, 134
Moremi Game Reserve, 60
Mount Karisoke Research Centre, 101–3
Mozambique, 25, 62, 94, 95, 130, 147,
  173, 196, 197
Mugabe, Robert, 103
Mursi, 58–9
Myanmar, 27
Myers, Norman, 2

Nam Theum 2 Dam, 71–2
Namibia, 25, 95, 121, 122–3, 133, 140
national parks,
  community-based approaches, 97–8,
    133, 221–3; *see also* CBNRM
  creation of new national parks, 5,
    74–5, 87
  elephant populations, 129, 131, 141,
    144
  fortress conservation, 9, 44, 77, 98,
    103, 111, 219, 221–3
  models of, 4–5, 149
  and NGOs, 184
  removal of communities, 56–62
  tourism, 152, 223
  see also protected areas
National Wildlife Crime Unit
  (NWCU), 33, 218
Nechisar National Park, 58–9
Nepal, 27, 47
new wars, 164, 173–4,
Nicaragua, 201
Ngorongoro Crater, 12, 57, 152
Nkunda, Laurent, 161–2

non-governmental organizations
  (NGOs), 1–4, 8, 27, 49, 61, 83,
    97, 184, 206, 219
  funding, 66–76, 138
  growth in power, 64–6, 105–6
  knowledge brokers, 67–8, 138, 145,
    221
  role in the ivory trade ban, 126–8,
    132, 136–9
North Korea, 27

Oates, John Frederick, 99
Okavango Delta, 60, 79–80
Olindo, Perez, 135–6
Operation Charm, 33–4
Operation Ribbon, 22
Operation Uhai, 82
Organisation of African Unity, 165
Organised crime, 17, 22, 34, 81, 118,
  188, 211–15, 217
Oxfam, 164, 165

Parc National des Volcans, 102
Partnership for Action Against Wildlife
  Crime (PAW), 33, 35
people trafficking, 39, 118
poaching, 68, 86, 88–90, 94–6, 223
  anti poaching, 10, 81, 96–111, 221,
    219
  definitions, 81, 82–93, 109
  shoot to kill, 77, 81, 82, 96, 99,
    103–4, 106, 110, 135, 221
  sponsorship of rangers, 83, 105
  see also hunting
precautionary principle, 49
protected areas, 206
  displacement, 53–63, 71–6, 222
  see also national parks

Qit Minerals Madagascar (QMM), 6–7,
  174

Ravalomanana, Marc, 175
Renamo, 95, 103
rhinos, 43, 114–23
  black rhinos, 4, 115, 122
  dehorning, 120
  Indian one horned rhino, 115, 116
  Javan rhinos 115
  poaching, 82
  rhino horn farming, 115, 121
  sport hunting, 121–3
  Sumatran rhinos, 115
  trade ban, 116, 119–23

trade in rhino horn, 10, 13, 18, 50, 218
uses of rhino horn, 116–19
value of rhino horn, 120
white rhinos, 3, 115, 122
Rio Tinto, 6–7, 68, 174
Roosevelt, Theodore, 84
rosewood 194–6; *see also* wooden carvings,
Royal Society, 5
Royal Society for the Prevention of Cruelty to Animals (RSPCA), 33
Royal Society for the Protection of Birds (RSPB), 3, 5
Russia, 27, 28; *see also* Soviet Union
Russian mafia, 22
Rwanda, 101–3, 156, 159, 216
    Interahamwe, 166, 167–8
    Rwandan genocide, 162, 165–7, 173
    Rwandan Patriotic Front (RPF), 166, 167
Rwanda Tourism Authority, 102

Safari Club International, 35
Save the Rhino International, 115
seahorses, 34, 45–6, 190
seashells, 189–90, 192–3
Selous, Frederick Courteney, 84
Serengeti, 12
Shell, 68
Sierra Club, 66
Smithsonian Institute, 7
snow leopards, 28
Society for the Preservation of the Wild Fauna of the Empire (SPFE), 84
Somalia, 20
South Africa, 8, 74,194
    abalone trade, 23–6
    elephant populations, 130, 131
    protected areas, 54, 62, 131
    role in ivory debate, 140, 144–6
    role in regional conflict, 94–6, 106,
    South African Defence Force (SADF), 95, 107
    trading wildife, 121
    trophy hunting, 122; *see also* hunting
    see also Kruger National Park
Soviet Union, 22, 44, 113, 173; *see also* Russia
Sri Lanka, 45, 188, 190
Sudan, 107

Taiwan, 20, 28, 117, 119
Tanzania, 12, 20, 57, 126, 135, 166, 197, 216
Terborgh, John, 98
Thailand, 8, 15, 27, 192, 202–5
The Nature Conservancy (TNC), 3, 53, 66, 71–3
Tibet, 28, 89–90
tigers, 4, 43
    captive tigers in USA, 29
    decline in numbers, 26, 27, 88
    National Tiger Conservation Authority, 89
    poaching, 68, 81, 86, 88–90
    Project Tiger, 27
    Save the Tiger Fund, 3, 30
    tiger farms (China), 30–1
    trade in tiger parts, 10, 18, 26–31, 50, 89–90
tortoises, 189
tourism, 108, 223
    beach building, 207–10, 215, 222
    photographing captive animals, 202–3
    safari tourism, 79–80, 84–5, 87, 114, 129, 143, 146, 152–3, 205, 215, 216, 222
    sand piracy, 207, 210
    working elephants, 203–205
Traditional Chinese Medicine (TCM), 18, 28, 29–31, 33–4, 45, 88, 90, 116–19
TRAFFIC, 18–19, 21, 24, 26, 27, 28, 29, 51, 89, 92, 192, 194
TRAFFIC-Asia, 19, 32
TRAFFIC-International, 9
Transfrontier Conservation Areas (TFCAs), 62
Triads, 35, 118
Turkey, 27
turtles
    egg collection, 190
    nest disturbance, 208–9, 215
    turtleshell, 188, 191–3

Uganda, 155–6, 159, 165, 216; *see also* Bwindi Impenetrable Forest
UN Convention Against Transnational Organised Crime, 88
UN Office on Drugs and Crime (UNODC), 88
UNITA, 95, 173; *see also* Angola
United Kingdom (UK), 19, 31, 36, 43, 92, 134

United Nations (UN), 65, 88, 125, 160, 166, 169
United States Agency for International Development (USAID), 73, 74, 175
United States of America (USA), 19, 36, 43, 45, 75
US Bureau for International Narcotics and Law Enforcement Affairs, 214
US Endangered Species Act, 51
US Fish and Wildlife Service, 30
US Rhinoceros and Tiger Conservation Act, 119
US State Department, 18, 214

Vietnam, 27, 28, 45
Virungas National Park, 101, 102, 159, 161–2

Walmart, 69
war in Central Africa, 165–74
    proxy rebel movements, 167
    role of Uganda and Rwanda, 167–70,
wilderness, 51, 53–63, 131, 206, 221, 223
Wildlife Conservation Society, 3, 6, 66, 70, 75, 80, 163, 181
Wildlife Direct, 100, 104–5, 160–2; see also Leakey, Dr Richard
Wildlife Protection Society of India, 89
wildlife trade, 17–19, 218
    drivers of, 8–10, 13, 17, 19, 36, 43, 117, 168–9, 218–19
    enforcement, 18, 31–7, 44, 51, 194, 218
    international co-operation, 45–6, 218
    Internet sales, 36–7

role of criminal networks, 34, 35, 37, 43, 118, 217–18
role in extinction, 18, 46, 223
tourism, 187, 189, 222
trade in marine life, 20–6
value of, 9, 17–18, 217
see also abalone; caviar; cockling; elephants; ivory; organised crime; rhinos; seahorses; tigers; wooden carvings
Wilson, E. O., 2, 52
wooden carvings, 188, 194–8
World Bank, 6, 8, 67, 70, 72–3, 97, 134, 175, 176, 205, 222
World Parks Congress, 74
World Wide Fund for Nature (WWF), 1, 2, 4, 6, 19, 49, 53, 66, 73, 80, 89, 161, 177, 222
    and the ivory trade ban, 126–8, 137–9
    WWF-Africa, 138
    WWF-International, 65, 106, 137–8, 192
    WWF-Madagascar, 182
    WWF-Switzerland, 137
    WWF-UK, 2, 34, 138
    WWF-US, 3, 92, 128, 138

Yakuza, 35, 118
Yellowstone National Park, 4, 56–7
Yemen, 117
Yosemite National Park, 4, 56

Zimbabwe, 25, 62, 106, 120, 129, 130, 165, 168, 196
    lobbying for the ivory trade, 140–3
    war on poachers, 103–4